Methods of Presenting Fieldwork Data

**Peter St John and
Dave Richardson**

THE GEOGRAPHICAL ASSOCIATION

Acknowledgements

The authors would like to thank the following people:

Mr E E Jones, the Warden of Lancashire Field Study Centre, Hothersall Lodge, for his continual support and for supplying the photographs;

Mrs E Kenny, for typing, retyping and retyping the various drafts, always with a smile;

J M Fallas for the cartoons;

the staff and students of many Lancashire schools and colleges who have indicated the need for just such a manual and who have fully field-tested the contents over a number of years.

ISBN 0 948512 16 4

Cover design: Chris Hand
Cover photograph: A level students carrying out a river study on Langden Beck, Forest of Bowland, Lancashire *Photo: Dave Richardson*
Printed and bound in England by Stephen Austin & Sons Limited

First published 1989
Reprinted 1995
Reprinted April 1996

The views expressed in this publication are those of the authors and do not necessarily represent those of the Geographical Association.

Published by the Geographical Association, 343 Fulwood Road, Sheffield S10 3BP. The Publications Officer of the GA would be happy to hear from other potential authors who have ideas for geography books. You may contact the Publications Officer via the GA at the address above.

The Geographical Association is a registered charity: no 313129

Contents

Introduction

Fieldwork plays an important part in any geography, biology or environmental studies course and selecting the most appropriate method of presenting the field data is as important as collecting the data itself. There are many texts that aid students in the setting up of fieldwork investigations and that guide them through the various techniques of data collection. This manual is designed to complement such publications by:

a. enabling fieldworkers to use their imagination in selecting the most appropriate techniques for the type of data they have collected;

b. illustrating the wide range of graphical and cartographical techniques available to them; and

c. making students confident in the use of these different and often more sophisticated techniques, many of which appear as data response questions in their final examinations.

The table below illustrates the various techniques

Type of Information	Technique	Page(s)
1. Introducing the reader to the subject matter or area	Field sketches Tabulation Base maps Flow diagrams	6 8 44 10
2. The organisation of raw data into a manageable form	Tabulation Ways of classifying data Use of logarithms	8 11 34
3. Representation of sequential data that changes over time	Line graphs Circular graphs Pictograms	12 31 14
4. Observed data at specific sites or locations that have definite component categories	Barcharts and histograms Pyramid graphs Multiple and composite graphs Mirror graphs Reverse bars Pie graphs	16 18 20 19 33 22
5. Representing connections between two sets of data	Scattergraphs/correlation graphs Mirror graphs	24 19
6. Representing data that shows a definite orientation	Rose diagrams Polar co-ordinates	28 32
7. Data that is composed of a number of elements that go to make up a total of 100%	Triangular graphs (3 elements) Composite bargraphs Block graphs	30 21 43
8. Where measurements of side views have been taken	Profiles Cross sections	36 37
9. Where data has been collected either continuously or at intervals along a sample line (a transect)	Scattergraphs Profiles and cross sections Mapping and divided bars Kite diagrams Block graphs	24 36 41 42 43
10. Data has been collected to show spatial variation	Base maps which then use the following: Dot maps Symbols and proportional symbols Choropleth and isopleths Location quotient	 44 46 50 54
11. Where data has been collected to show spatial variation of movements and flows	Composite bars Flow lines Desire lines	55 56 58

described in this booklet and shows when they may be used (please note: the authors consider this publication as a manual into which students may 'dip' for inspiration rather than the definitive document on the subject). The exact use of each technique depends on the precise nature of the data.

Wherever possible the presentation of each technique has been standardised onto double facing pages, this should make the manual easy to use as all the information needed is available without having to turn any pages. Each technique appears whenever possible under the standard headings of When to use, Examples, Method of construction and Worked examples. In some cases an additional section on drawbacks/criticisms of the method may be included.

The authors have also included a Glossary of terms which explains the meaning of terms used throughout the manual.

Those students who wish to further their fieldwork analysis by the use of simple but relevant statistical techniques should consult the 'sister' manual to this one entitled *Statistical Methods of Analysing Fieldwork Data* (also to be published by the GA).

Field Sketches

These are exactly what the title suggests, a sketch made in the field showing a landform or feature of interest. A good photograph of the same feature may be useful but it can never be a real substitute for a well drawn and labelled field sketch. The reasons for this are as follows:

a) Making a field sketch will force the observer to look more closely at the feature and to record his observations carefully at that particular time and place.

b) A field sketch allows the observer to select, emphasise and even omit whichever details he or she wishes. It is possible for a photograph to show too much detail. A good field sketch should omit the irrelevant and emphasise the relevant.

c) It is easier to use a field sketch than a photograph as a base for labels. These labels may convey concepts such as location, explanation, that no other method (graph, map, transect) is able to communicate. As such, field sketches are a major method of recording data in the field.

d) At best field sketches create a summary of the observations made at that time in a dramatic and vivid way. They can add a visual impact to any fieldwork project.

The main problem with field sketches is that they must look like the feature depicted. This is no easy matter as very few landscapes are simple and not all people are natural artists. It is argued that with the help of the following tips and plenty of practice, either in the field or from pictures/slides, most pupils can reach an acceptable standard of sketching.

i) The most common error in sketching is to over-exaggerate the vertical, for example hills are commonly drawn too high. This can be overcome by dividing the page into three equal parts by drawing two lines across the page at equal intervals. Use these as guidelines. Squares may be used instead.

ii) Decide on the boundaries of what you want to draw. A natural feature such as a row of trees, spur of land, etc. can often be used. Sometimes a cardboard frame held at arm's length can be used to define the area and give an idea of scale. If a transparent sheet is put over the middle of the frame then it is possible to trace the major features on to it, using a felt tip pen. This is then later transferred onto the paper as a base.

iii) Sketch on the horizon and then the other major features. Hold the pencil at arm's length and using it as a marker work out the rough proportions of these major features. If this part is done well the rest will follow more easily.

iv) Fill in section by section the detail required omitting the irrelevant. If grid squares have been used then each is completed in turn until a complete picture has been built up.
Shading and lines can be used to great effect in showing slope and angle. Other symbols may be used for other features such as woodlands, fields, water, etc. Not every minute detail is needed only enough detail to illustrate the point you are trying to make. If possible try to give an idea of scale.

v) Give the sketch a full and detailed title. This tells the reader instantly the location and what you are trying to show. The title should include where the sketch was taken from (Grid references), the direction the observer was facing and what he or she is trying to show (land use, physical features, settlement patterns, etc.).

vi) All features must be fully labelled. If possible label different features such as hills, settlements, in a different colour or different type of print. These help to orientate the sketch for the reader.

It is possible to produce sketch maps of such features as rivers, glacial features, and beach deposits using the same principles as those outlined above.

Worked example

Photograph

Field Sketch

Field sketch of Langden Brook looking downstream from G.R. 603504 showing the main valley features.

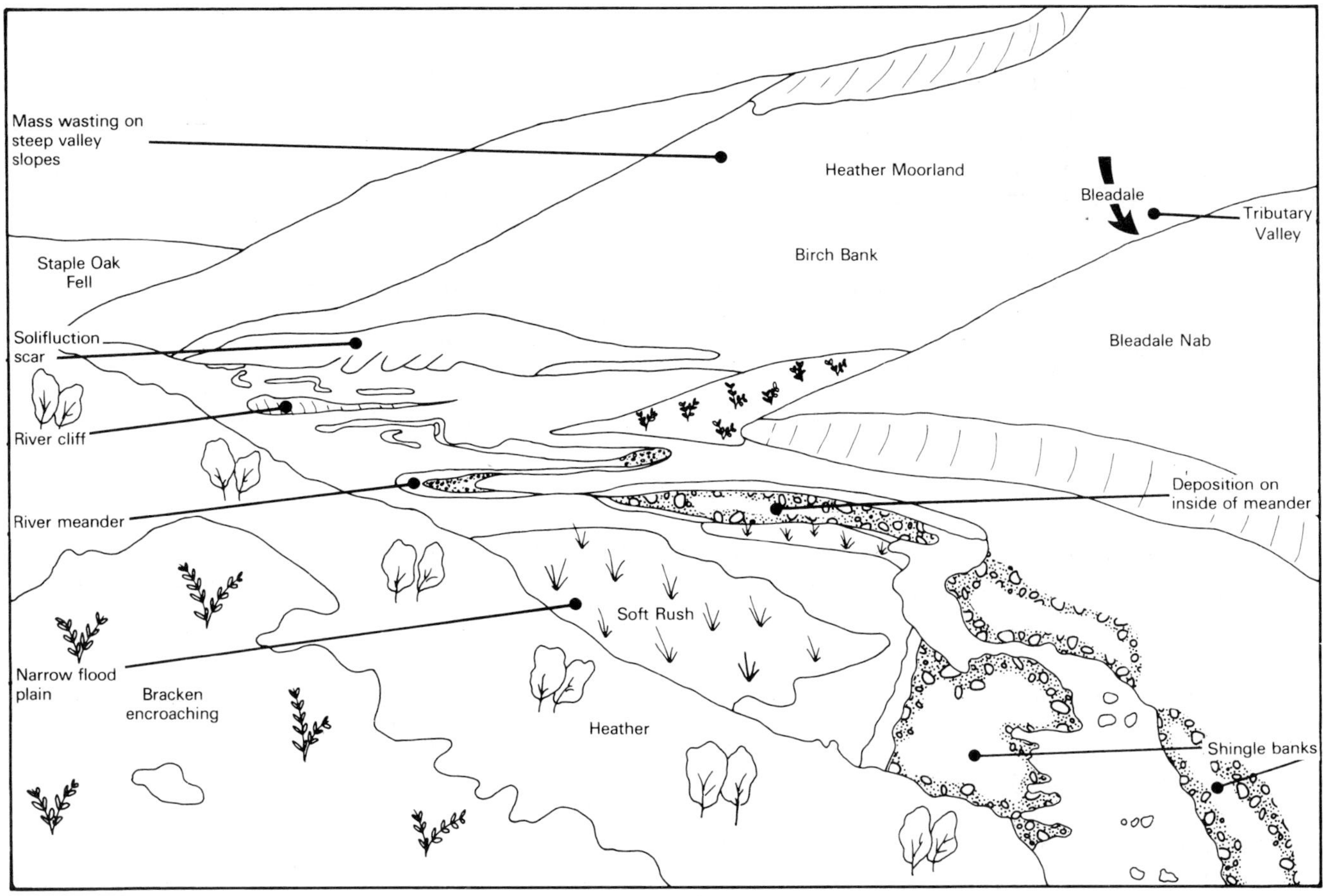

Mass wasting on steep valley slopes

Heather Moorland

Bleadale

Tributary Valley

Staple Oak Fell

Birch Bank

Bleadale Nab

Solifluction scar

River cliff

River meander

Deposition on inside of meander

Soft Rush

Narrow flood plain

Bracken encroaching

Heather

Shingle banks

Tabulation

It is important to make clear and concise tables both in the field and as an integral part of the fieldwork project. They show the reader where much of the information has come from and how it was collected. A well structured table can be in many cases a worthwhile method of representing data. There are no set rules for constructing tables as the nature of the data tends to dictate the nature of the table. The following are therefore suggestions.

1. All tables must have a title and must be fully labelled. This makes it obvious to the reader what the table is trying to show and it should therefore be unnecessary for him or her to search for further explanations in the script.

2. Wherever possible show the 'raw' data, that is data collected in the field. This should be the first table in that particular section. It is equally possible to put this into an appendix and refer to it in the text. Values manipulated from this raw data (percentages, means, etc.) should then be put into a further set of secondary tables.

3. It is a common fault to try to put too much information onto one table. It is better to put summaries onto separate tables.

4. Include totals for each row and column *but* only if relevant. A grand total may be included in the bottom right hand corner as a final check by addition of all the rows and columns.

5. If the type of totals are different (partial, accumulative, grand, etc.) make sure they are labelled as such especially if they appear on the same table.

6. If numbers have been converted in some way (percentages, formulae, etc.) show clearly what the conversions are and where the original data has come from.

7. Units of measurement must be included at all times.

8. It may be useful to leave room for later additions. It is generally easier to extend the number of rows rather than the number of columns.

The following worked examples hopefully will illustrate some of the different types of tables that may be used.

Worked examples

Stream channel variables (normal flow) for Langden Beck, Lancashire

Channel Variables	Site Number									
	1	2	3	4	5	6	7	8	9	10
(1) Distance downstream in m.	200	1,100	1,900	2,400	3,100	3,900	4,200	5,300	5,900	6,500
(2) Long profile gradient in degrees	6.2	5.1	3.1	4.6	2.5	1.5	1.8	0.4	1.3	0.5
(3) Mean velocity in m/sec	0.78	0.62	0.54	0.91	0.89	0.99	1.2	1.34	1.21	1.45
(4) Cross section area — m^2	0.12	0.34	0.48	0.51	0.62	1.21	1.45	1.56	1.85	1.89
(5) Discharge in Cumecs (3 × 4)	0.09	0.21	0.26	0.46	0.55	1.19	1.74	2.09	1.89	2.74
(6) Wetted perimeter in ms	0.86	0.94	1.12	1.63	1.97	2.46	3.14	4.92	5.67	4.81
(7) Hydraulic radius (4 ÷ 6)	0.14	0.36	0.43	0.31	0.31	0.49	0.46	0.32	0.33	0.39
(8) Mean stone size (a axis) in mm.	289	198	201	189	134	145	101	86	92	67
(9) Sphericity index (after Krumbein)	0.36	0.32	0.56	0.65	0.54	0.64	0.72	0.53	0.78	0.75

Spearman Rank Correlation Coefficient
Distance downstream is correlated against long profile gradient using data from example 1.

Variable A (Distance down-stream in m)	Rank	Variable B (Long profile gradient)	Rank	Difference in Ranks (d)	d^2
200	10	6.2	1	9	81
1,100	9	5.1	2	7	49
1,900	8	3.1	4	4	16
2,400	7	4.6	3	4	16
3,100	6	2.5	5	1	1
3,900	5	1.5	7	−2	4
4,200	4	1.8	6	−2	4
5,300	3	0.4	10	−7	49
5,900	2	1.3	8	−6	36
6,500	1	0.5	9	−8	64
				Total (Σ) d^2 = 320	

Spearman rank (rho) $= 1 - \left(\dfrac{6.\Sigma d^2}{n^3 - n}\right)$ where n = number of paired observations[3].

$\therefore$ rho $= 1 - \left(\dfrac{6 \times 320}{1000 - 10}\right) = 1 - \left(\dfrac{1920}{990}\right) = \underline{\underline{-0.939}}$

There are (n) degrees of freedom (10).

Longridge Census Data, 1961 — age and marital status

Age last birthday	Males			Females		
	Total	Single	Married	Total	Single	Married
0 − 4	190	190	—	206	206	—
5 − 9	214	214	—	209	209	—
10 − 14	187	187	—	198	198	—
15 − 19	174	173	1	151	141	10
20 − 24	122	78	44	142	48	94
25 − 29	147	34	113	132	12	120
30 − 34	154	22	131	150	10	140
35 − 39	149	16	130	150	14	133
40 − 44	117	14	103	115	16	96
45 − 49	155	11	144	171	15	145
50 − 54	154	17	129	180	18	143
55 − 59	148	11	133	165	33	106
60 − 64	118	10	103	163	23	100
65 − 69	79	9	60	119	25	61
70 − 74	63	2	51	105	24	36
75 − 79	33	1	20	64	16	17
80 − 84	6	—	6	27	2	3
85 − 89	5	—	3	13	3	1
90 − 94	1	—	1	—	—	—
95 − 100	—	—	—	—	—	—
Total	2226	989	1172	2460	1013	1205
Widowed	61			236		
Divorced	4			6		

Flow Diagrams (systems diagrams)

When to use

Many fieldwork projects may concern the investigation of a particular geographical or ecological process or system. It may be useful to show the system under investigation either by means of an introduction to the work or as a conclusion once the analysis has been completed. One of the best ways of doing this graphically is by using a flow diagram (sometimes called systems diagram).

Examples

The following systems are examples that may be shown in a diagrammatical form using this method — slope, fluvial, marine and glacial processes, plant successions, woodland ecology, agricultural, industrial, settlement and transport systems.

Method of construction

Most modern textbooks contain systems diagrams but it is far better for the fieldworker to construct his or her own diagrams as they relate to their own specific and unique study.

i. Identify each of the stages in the system and where they appear in the sequence. Place each one inside a frame or 'box'.

ii. Arrows are constructed to indicate where a relationship or flow exists between each of the stages (boxes). The ordering of the stages is therefore only achieved by careful forethought or even trial and error.

Note. The two stages just described will give a simple diagram as shown in the worked example below but it is possible to construct more complex diagrams called 'constellation diagrams'. As these are much more complex and specific in nature, a brief description and a worked example are included in the Appendix (see page 60).

Worked example

Formation and development of a limestone pavement

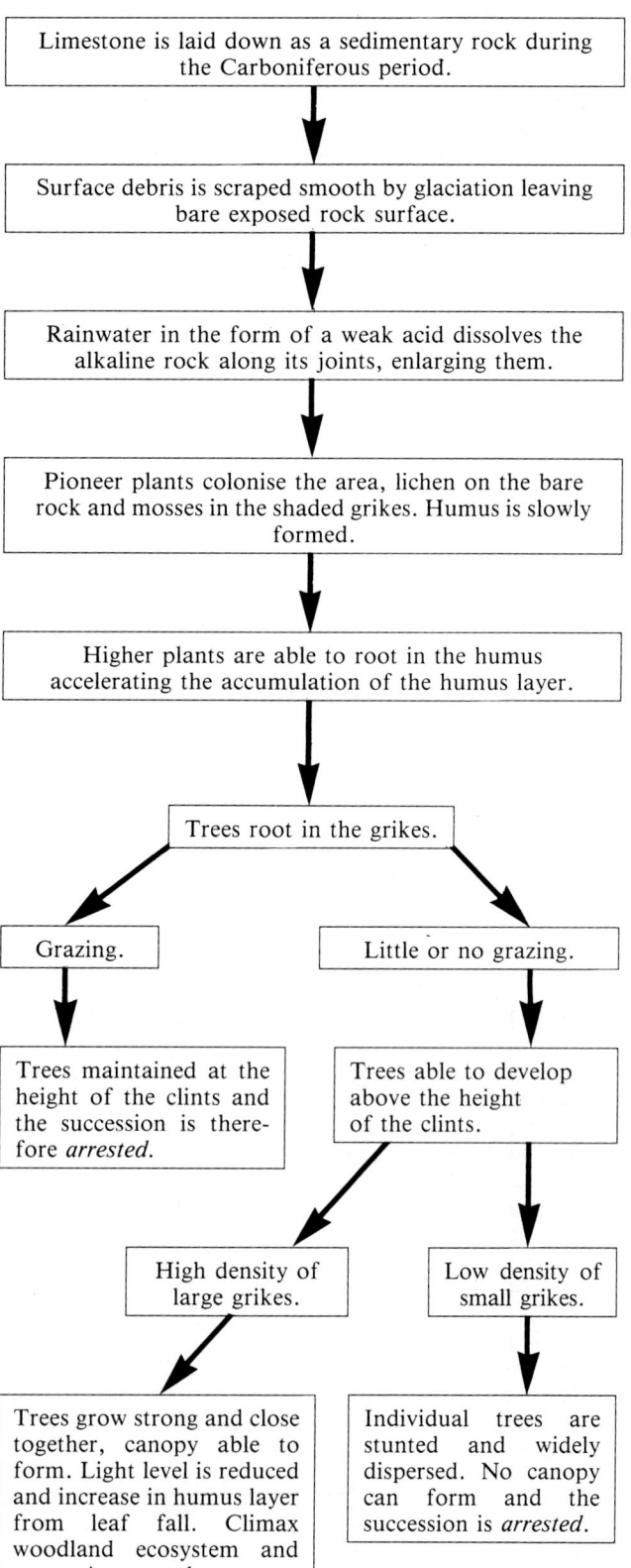

Different Ways of Classifying Fieldwork Data

In many cases data that has been collected in the field, usually in preparation for analysis as part of a fieldwork investigation, has to be processed before it can be used. Many of the techniques in this booklet need the data classified before it can be graphed or mapped, for example choropleth mapping, flow lines, etc. There are, however, several ways of putting 'raw' data into groups. The first step is to look at how the individual observations are spread across the range. This is best achieved by constructing a **dispersal diagram** as shown below.

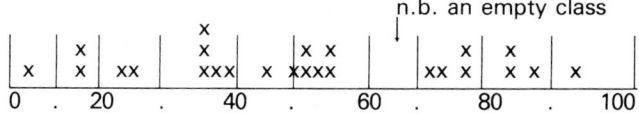

Observed values

a. Fixed intervals. The data range is divided into regular size classes (0 – 9.9, 10 – 19.9, etc). The frequency of occurrences in each of the classes is then noted.

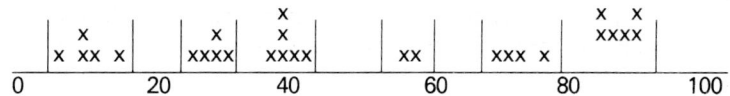

This is the most common method used but if the points are unevenly distributed it can lead to some misleading representations (this is especially true of choropleth mapping).

b. Percentile groupings. Groups are defined so that each contains an **equal number of points**. For example, if there are 25 observations and 5 categories then the first 5 points are placed in class 1, the second in class 2 and so on.

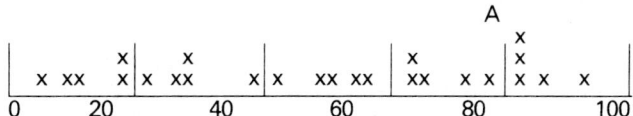

This is an improvement on the first method as it overcomes the problems of the data being unevenly distributed but it can give a misleading impression by artificially dividing up clearly related clusters of points, for example as at A on the diagram.

c. Arbitrary method. This is sometimes called the natural groupings method. Divisions are drawn in by eye according to natural clusters of points.

This method is useful when choropleth mapping as the final map gives a clear indication of reality but it may not be suitable for other techniques as the classes may be of uneven sizes and contain unequal numbers of data points.

d. Standard deviation classes. Descriptive statistics can be used in defining the classes. A data set may be divided on the basis of its mean and standard deviations. Classes can be used which represent -3, -2, -1, $+1$, $+2$, $+3$ standard deviation.

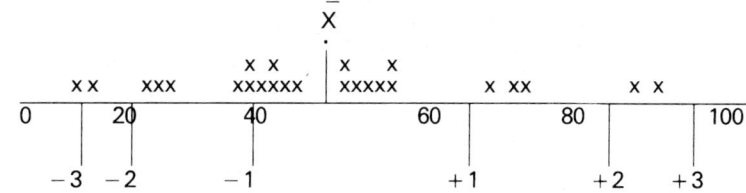

This method brings out which areas are exceptional to the main body of the data but it does have the disadvantage, in normally distributed data, of placing 68% of all the values in the two categories of $+1$ and -1 standard deviations.

Line Graphs

When to use

Line graphs are one of the most common ways of displaying results but in the past they have often been used where other techniques may have been more suitable. The line graph is most effective when the measurements to be shown are continuous. This makes the results sequential. The resulting line will show high and low values as 'peaks' and 'troughs'. It will also give a visual impression of the actual rate of change over time and/or space.

Example

Against time — temperature, air pressure oxygen/ nutrient content of a pond, traffic flows, population changes, farming and industrial output.

Method of construction

i.   Two axes are drawn (horizontal and vertical).

ii. The time intervals are put across the bottom (the horizontal axis).

iii. The scale for the observed measurements is labelled up the side (the vertical axis). This scale does not always have to start at 0. Choose an appropriate scale that will accurately reflect the nature of your results.

iv. Individual points are plotted as small but clear crosses (×). Use a sharp pencil.

v.   Using a ruler and a sharp pencil, join up the points in the order they occurred over time. It is possible to put on more than one set of data (results) on the same axes as long as the scales are identical.

Different colours can be used or different types of line as shown below. The cross that represents the plotted value must always be visible through the line.

Suggested graph lines

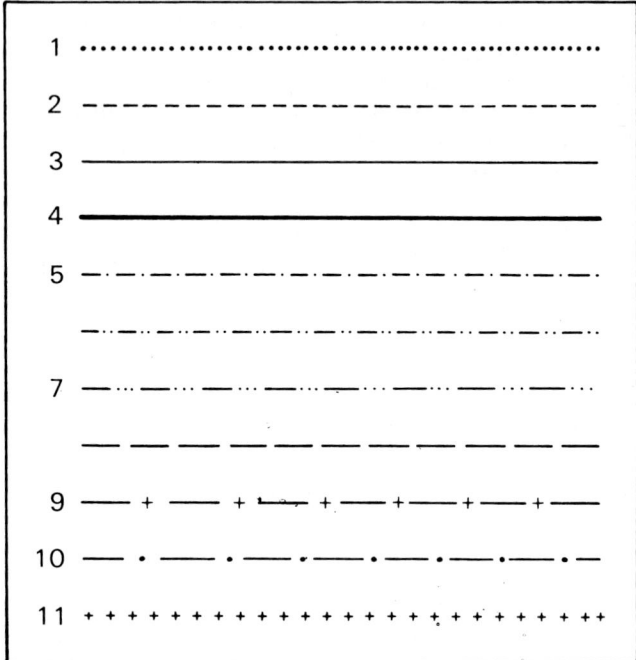

A word of caution

As already mentioned these graphs are often used to show or emphasise trends. Lines joining individual observed points can therefore be meaningful as in worked examples 1 and 2 *but* sometimes the observed trends are not meaningful, as in worked example 3.

Worked examples

Mean monthly temperatures — Perth, Australia

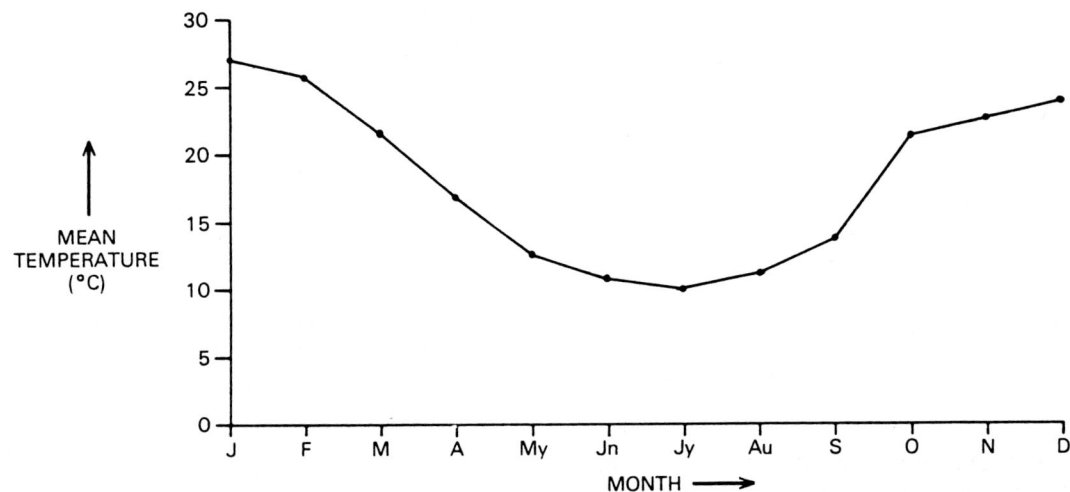

Population change for Longridge — 1950 to 1980

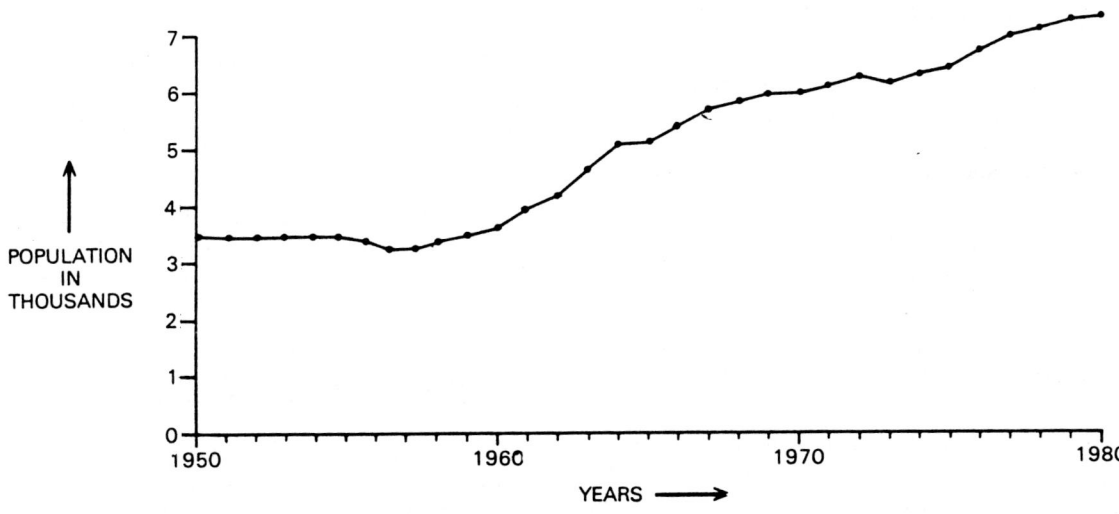

Cars sold in Longchester — 1945 to 1959

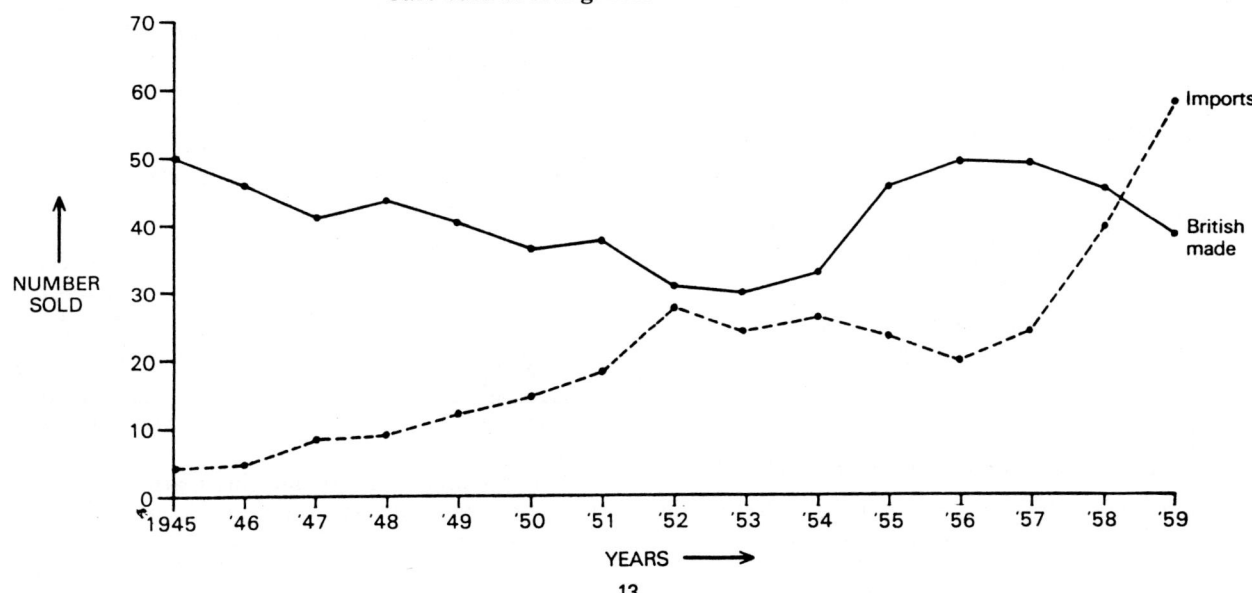

13

Pictograms

When to use

This method involves the use of *pictures* to show the observed results. There are two ways of using these pictures and these are shown below. This method is less accurate than line graphs or bar charts, especially when large numbers are involved. Pictograms are, however, visually attractive and should be used when visual attractiveness or appeal is more important than the need for accuracy. Pictograms can also be used on maps (see the section on mapping).

Examples

Any data showing change over time or space which can be represented by a single, simple picture — see below.

Method of construction

With this particular technique there are two ways of constructing the diagrams. Each has its advantages and disadvantages.

Method 1

i. The time/space scale goes up the side (vertical axis).

ii. The same picture is *Drawn* repeatedly across the page. Each picture is always the *Same size*. The values of the observation is shown by the total number of pictures drawn.

iii. As always a key has been worked out beforehand and is then displayed somewhere near the graph.

Worked example

Pupils attending Hothersall Lodge Field Study Centre 1972 – 1976

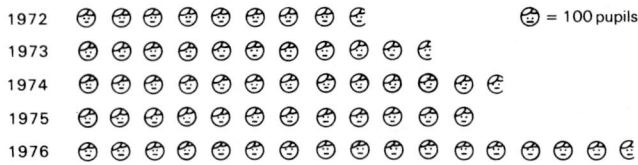

Method 2
This second method is much harder to construct.

i. A single line is drawn across the page with the time or space scale labelled (horizontal axis).

ii. A single *picture* is drawn for each observation. The value is indicated by *size* of the picture shown.

iii. The actual number/value is printed above the picture.

Worked example

Pupils attending Hothersall Lodge Field Study Centre 1972 – 1976

878	1066	1231	1201	1687

1972   1973   1974   1975   1976

As can be seen, as the observed numbers increase so do the heights of the individual figures, but this does not represent the data you want to display. This is explained in the next section.

Problems with this method

As already mentioned this method is not as accurate as others. In addition to this the *second method* is *not* recommended because

i. It is much harder to construct.

ii. It can be visually misleading. For example if the figures represented are being shown to double the change would probably be shown by doubling the height of the picture. If this is done then the width of the picture will also have doubled to keep the picture in proportion. This now means that the picture will be four times larger to look at and not doubled as desired.

It is possible to overcome this mathematically and this is discussed in the section on pie graphs (pages 22 and 23).

More worked examples

Numbers of cars sold in Longchester 1980 to 1985

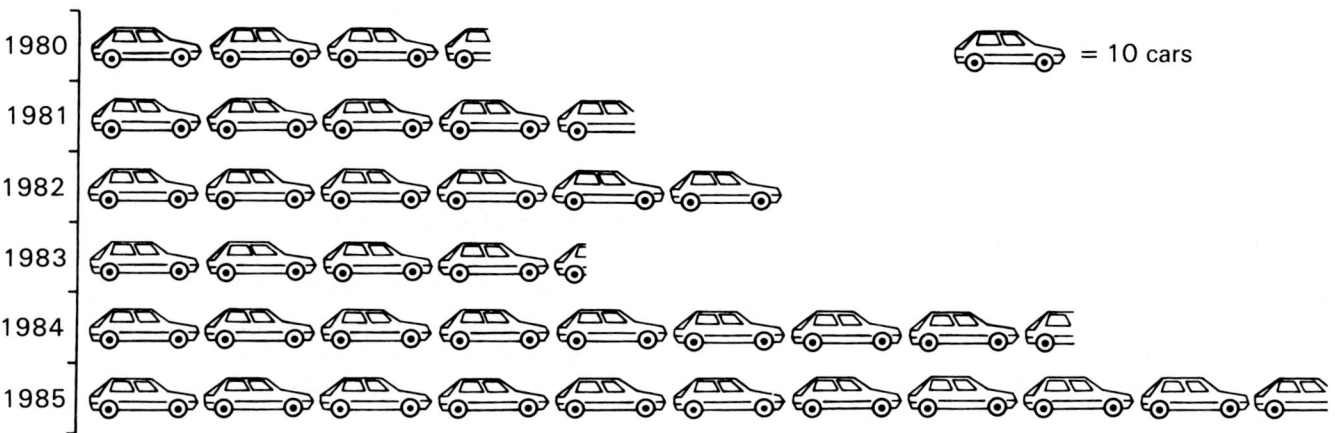

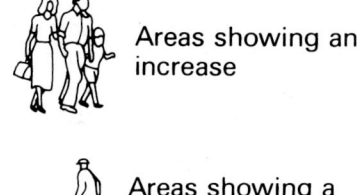 = 10 cars

Population changes in the districts of Lancashire 1971 to 1981

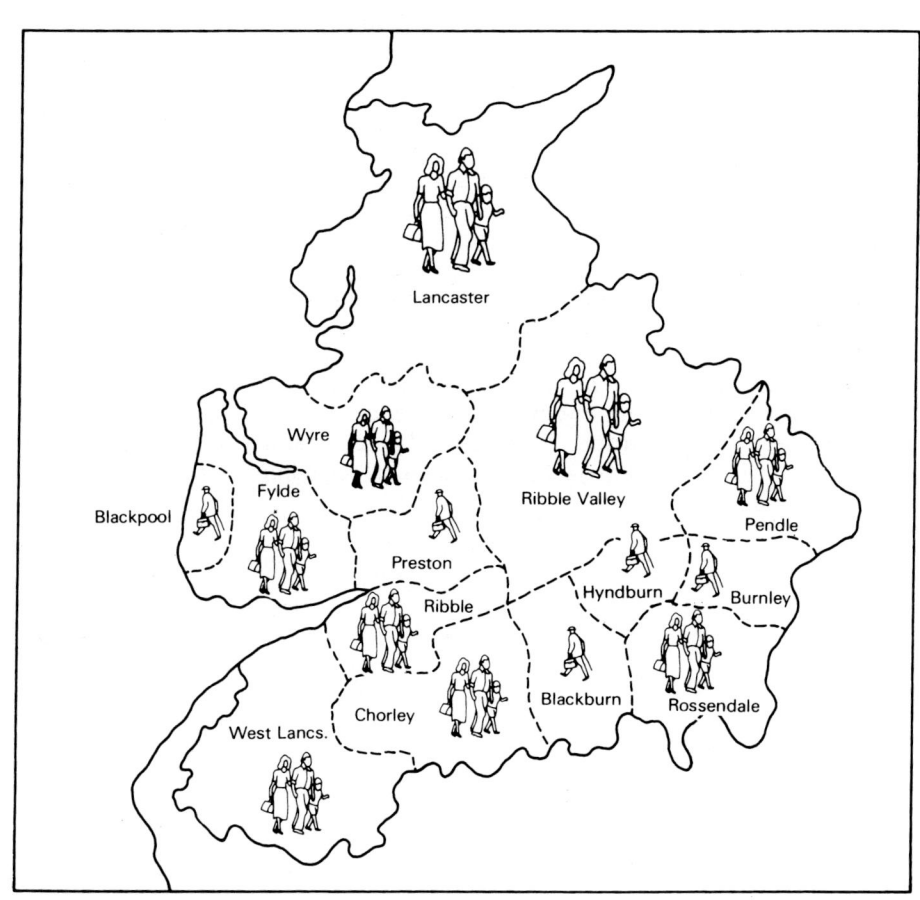

Areas showing an increase

Areas showing a decrease

Barcharts and Histograms

When to use

Barcharts and histograms are very similar visually as in both cases the data is shown as solid blocks or bars. Despite this there are some *very important differences*.

Barcharts

These have *one* quantitative scale. This is usually the vertical axis which is used to show the observed/measured data. This is often called the frequency. The *height* of each bar reflects the observed frequency. The horizontal axis does not have a numerical scale but has categories such as places, names, names of fauna and flora, etc.

Histograms

These have *two* quantitative scales. The vertical scale is the same as the barchart and shows frequency. The horizontal axis represents size classes or values, eg. pebble sizes, age categories, field size categories, etc. In the case of the histogram it is the *area* of the block that gives the observed frequency and not necessarily the height as in the case of the barchart. If equal intervals are used across the horizontal axis then this difference loses its importance but this is not always desirable. The problem of deciding on size classes is discussed in greater detail on page 11 (ways of classifying data). Both types of graph can be used on base maps (see the section on mapping).

Method of construction

i. Draw two axes in the usual way.

ii. The observed frequencies are put on the vertical axis. Start at 0 going up at regular intervals. The highest value on the scale should be just higher than the largest observed value.

iii. The horizontal axis is *either* the different categories in the case of the barchart *or* size class values in the case of the histogram.

iv. Label both axes carefully in ink.

v. Using a ruler and pencil draw in the values as rectangles or blocks. NB. Histogram blocks should be drawn connected together as the data is continuous. Barchart blocks should be drawn with a gap between each type as the data is not continuous (see worked examples). Remember these differences when working out the horizontal scale.

vi. Colour or shade in the blocks. Histograms show the frequency distribution of the whole sample and should therefore be shaded all the same. Barcharts show different, discrete categories and a different shade/colour may be used for each category.

The worked examples that follow will help to show the important differences between these two types of graph. Whichever type of graph you have make sure that you give it the correct name in the title.

Worked examples

1. Barcharts

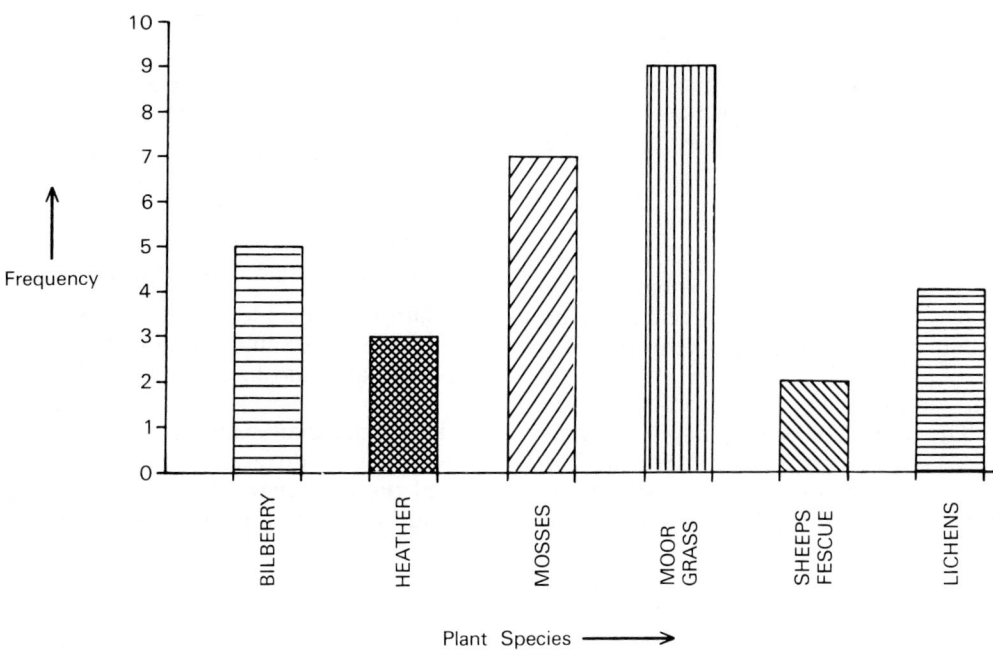

A barchart showing the results of a quadrat survey carried out on a stretch of open moorland in Yorkshire

Other examples

2. Histograms

Outputs (industrial, agricultural, etc.) from different countries, crops grown on a farm or from a region, tree type frequencies in a woodland, number and type of organisms in a river.

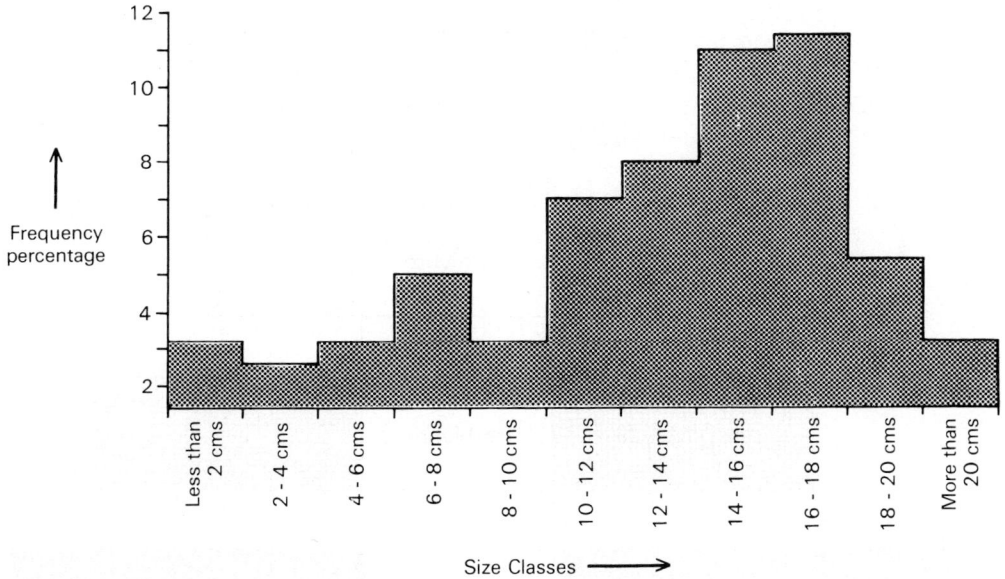

A histogram to show the size distribution of 100 pebbles measured on a stretch of mountain stream in North Wales

Other examples

Monthly rainfall totals at Hothersall Lodge for 1983, pebble shape indice distribution within a sediment sample, tree girth and height observations for a single species of tree.

As can be seen there are many examples and this type of graph is probably the most commonly used in fieldwork.

More Barcharts and Histograms

It is often possible to show more than one set of measurements (observations) on a single barchart and/or histogram. The following examples show the possibilities.

1. Pyramids

These are used when a *single* set of data is best shown in the form of a pyramid. The same numbers are used to draw blocks/bars on either side of a central axis.

i. The different categories are put in order up this central axis. The order can be very important and must be worked out very carefully in advance. For example the worked example below shows a food chain which involves a definite order starting at the base with primary producers and going through the chain up to the top carnivore.

ii. Frequencies/measured values are drawn twice across the bottom (horizontal) scale, one on either side of the central axis. Both start at 0 near the centre and then go up to a number just higher than the largest measured/observed value.

iii. Draw in the blocks/bars. Colour and shade the blocks. The same rules of spacing and shading for barcharts and histograms apply.

This particular example has spaced bars and each category has a different type of shading. It is therefore based on a barchart. If it were a histogram the bars would be continuous and shaded the same.

Worked example

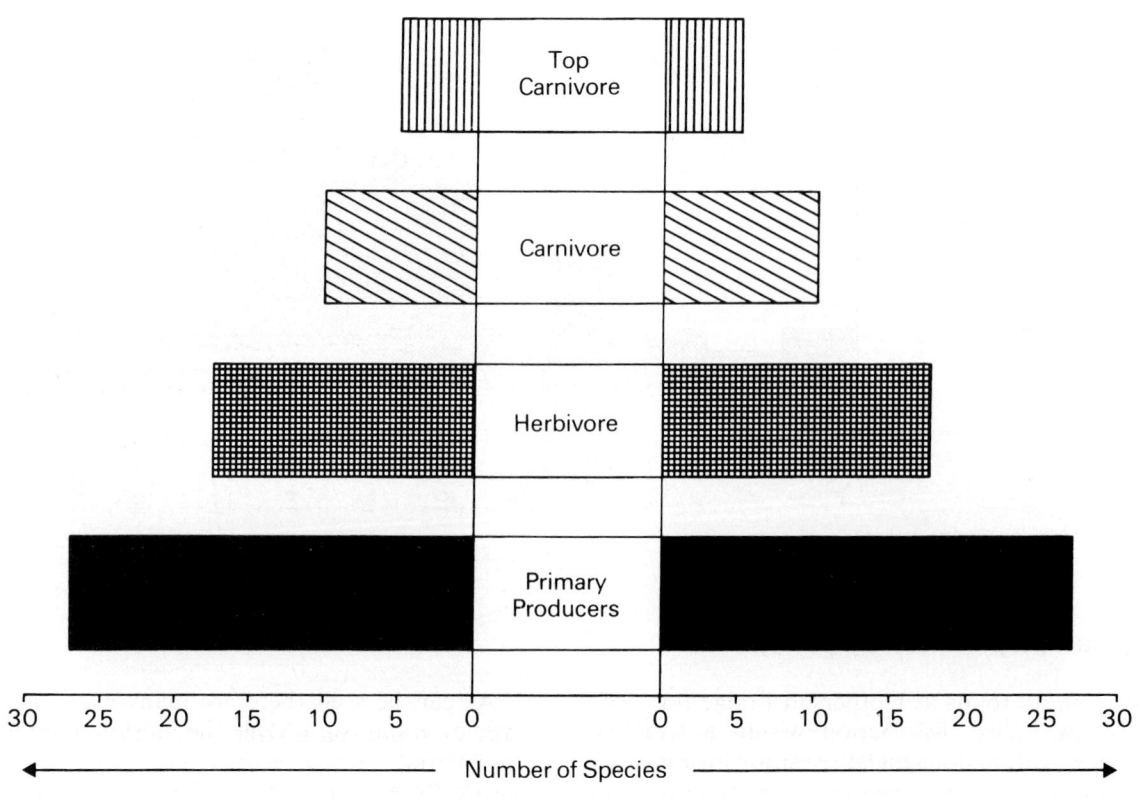

Population pyramid for a food web in a pond

18

2. Mirror graphs

These are very similar to the pyramids *but* they use two sets of measurements/data. The categories are labelled up the central axis; one set of data is drawn to the left of this central column whilst the second set of data is drawn to the right of the central column.

This method is ideal for comparing/contrasting two similar sets of data collected at two different locations. Once again the ordering of the categories up the central axis is very important. If organised properly this type of graph can be very attractive visually.

Worked example

**Vegetation survey on a podsol and a peat soil
(Bowland)**

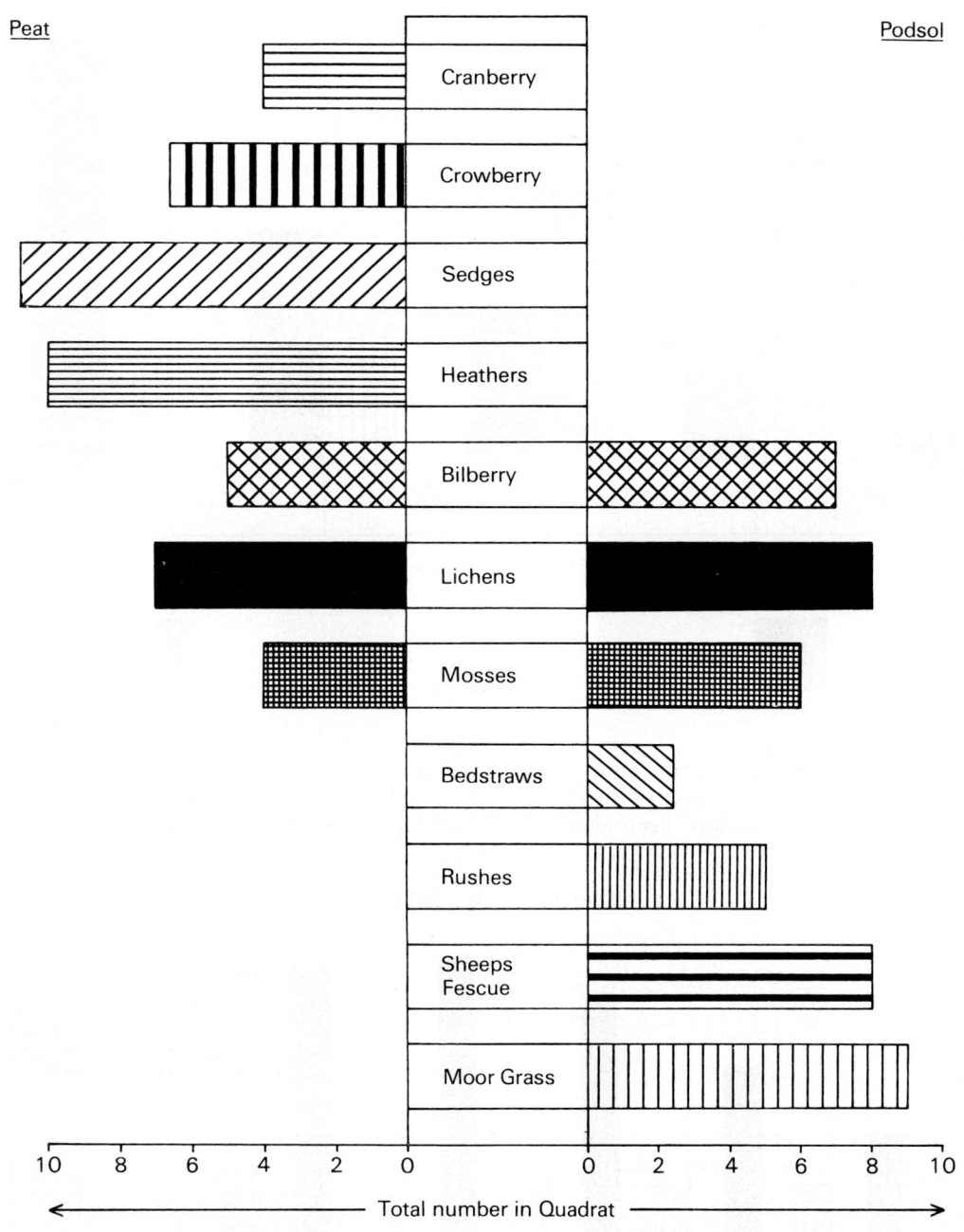

Again, based on barcharts, colours in this case would be better than the shading above. The following are further examples: pebble sizes or roundness at two different sites on a river; population pyramids (male and female treated separately); land use sampled at two different locations (highland and lowland).

3. Multiple bar charts/histograms

This is where several sets of data that have the same categories are shown on one graph. If there are only two sets of data use a mirrorgraph. If there are three sets of data use this method. If there are more than four sets of data draw separate graphs for each set of data as the multiple barcharts start to become complicated.

Each category will have several bars/blocks representing the different sets of data. The height of the bar stands for the value of each category. The horizontal scale is for the categories and the vertical scale is for the observed/measured values. Each set of data will have a different shading or colour. The worked examples below should make this clear.

Worked examples

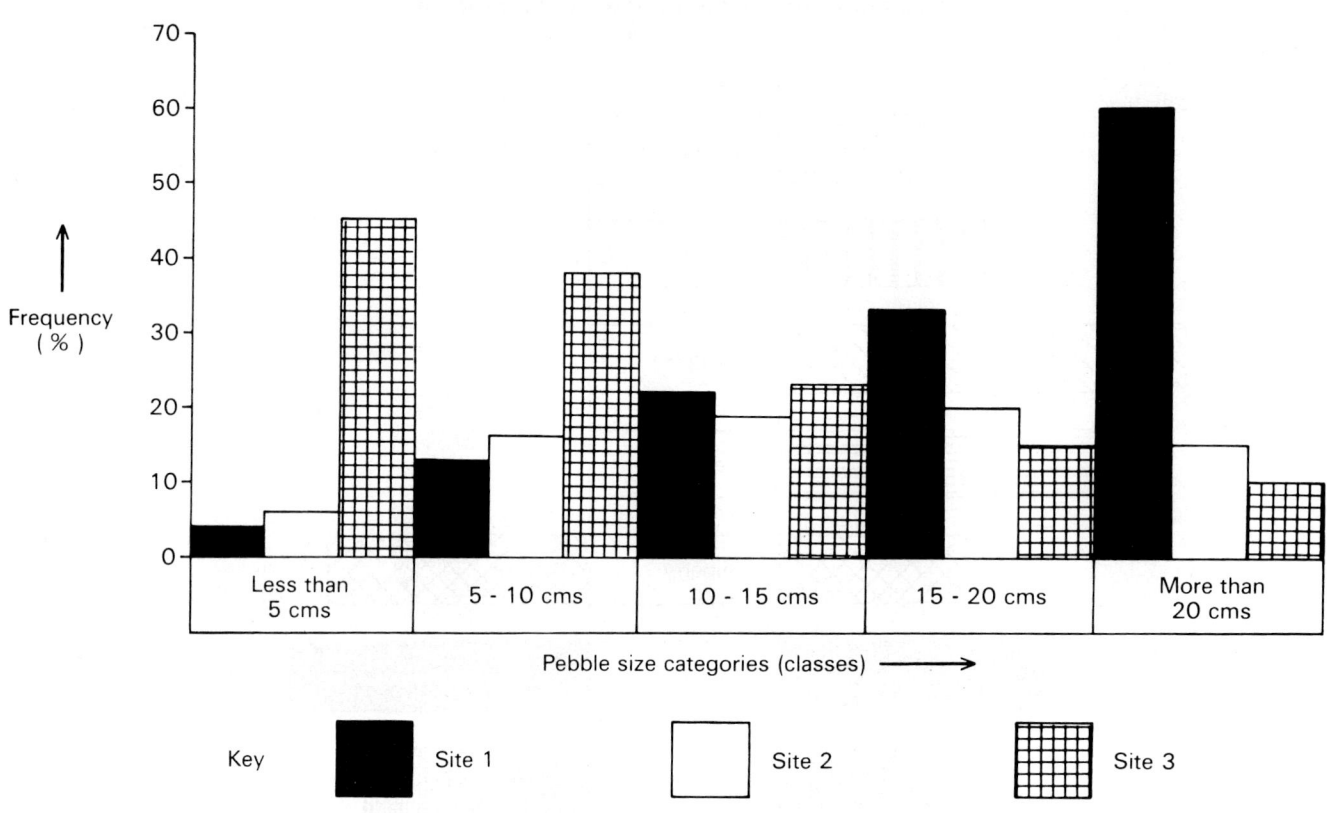

Stone survey at three locations along the long profile of a stream

The example above contains histogram data and therefore the categories are joined together as the data is continuous. Had the information been for a barchart then there would have to be a small gap between each of the categories (classes). The worked example below shows this sort of arrangement.

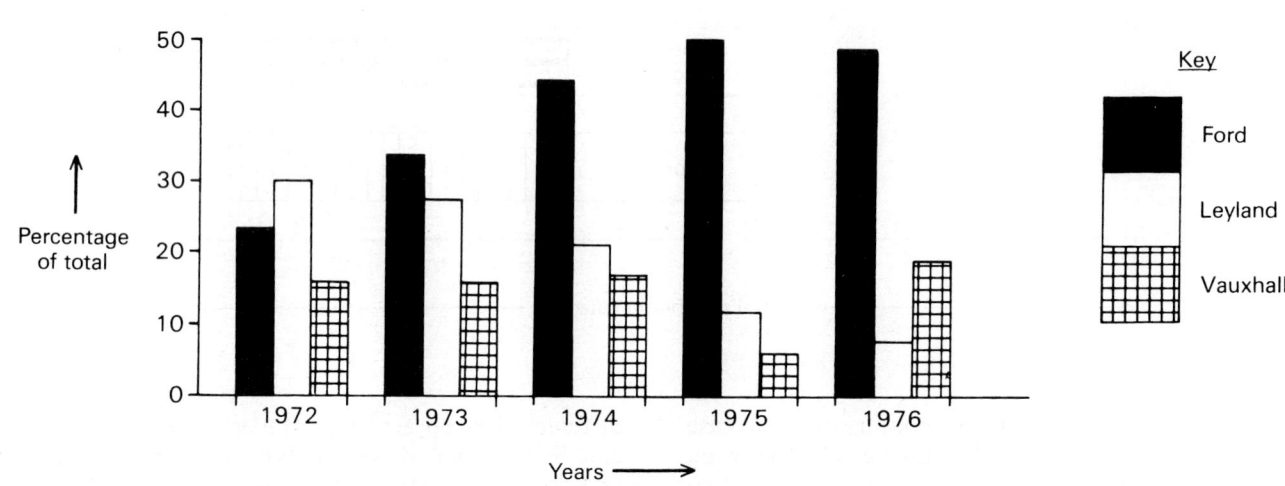

Cars sold in town 'A' (1972 – 1976)

4. Composite barcharts

This type of graph is usually used with barchart type data. Histogram data can sometimes be adapted and used but it is less common. The rules on when to use this type of graph are different from those governing multiple barcharts. Any number of sets of data can be shown on the graph but there should be no more than four or five different component categories within each of these sets of data.

There are two ways of drawing composite barcharts.

i. Each set of data is shown on the same graph as a single bar or block. Six sets of data equals six bars on the graph. The height of the bars represent the actual totals of each set of data. This way changes in the totals can be seen as changes in the length of the bars. The bars are then sub-divided into the different component categories according to size.

ii. If the total sizes are not important then the component categories of each set of data can be converted to a percentage of the whole. Each of the bars is therefore drawn the same size making a comparison of the different component categories more accurate and easier than in the first method.

The two worked examples below show these two methods using the same data as for the second example in the multiple barchart section. A table has been drawn to show how the percentages have been worked out, in Appendix 1.

Worked example

Cars sold in town 'A' (1972 – 1976)

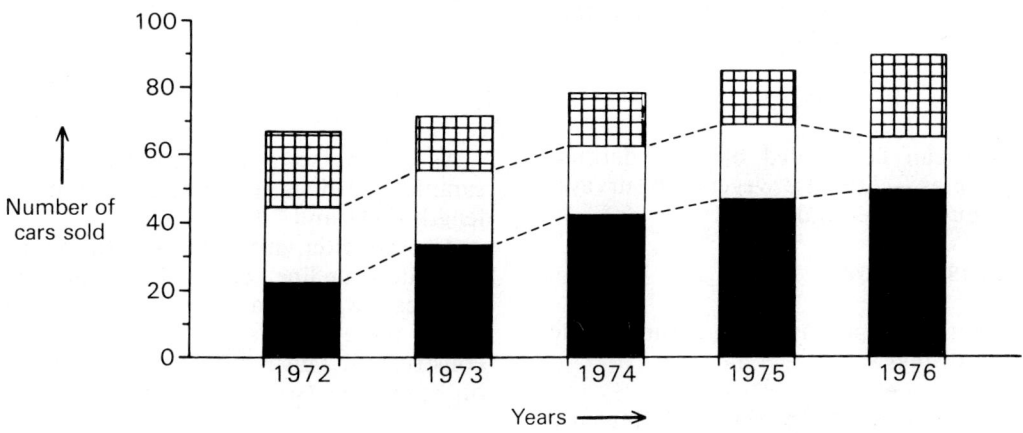

This graph shows the total changes from year to year but it is not easy to interpret changes in the component categories, ie. the makes of car. We therefore use a percentage composite barchart (as shown below) when this is necessary.

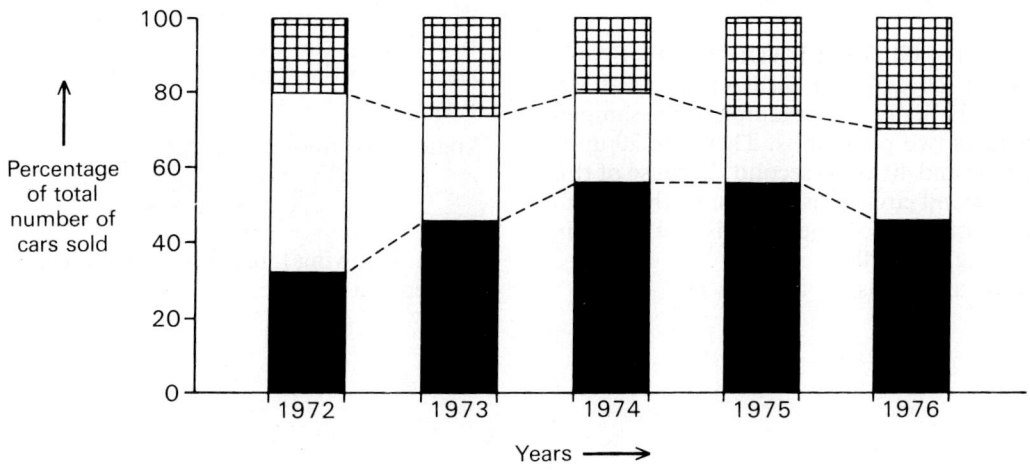

Pie Graphs

When to use

A pie graph is a circle that has been divided up into sections (like the slices of a cake or pie). These different sections are called the component categories. A finished pie graph gives a good impression of these different component categories and the relative proportions that go to make up the total, ie. the complete circle.

They tend to be used when there are four or more categories involved, which, as we have seen, tends to be too many for composite and multiple barcharts (pages 18 and 19). They are often used to compare two or more sets of data. If this is the case pie graphs that have more than five or six component categories become difficult to interpret. Pie graphs are commonly used to display results on base maps (see page 49).

Examples

Any sample that can be divided up into definite component categories — land use, vegetation surveys, traffic counts, housing ages and styles.

Method of construction

The particular method depends on the nature of the data you wish to display.

i. If the size of the different samples is different use **proportional circles**, eg. traffic counted at three sample points gave totals of 23, 45 and 13 vehicles.

ii. If the sizes of the different samples are the same *or* the component categories are to be expressed as a percentage of the total then draw all the circles the same size, ie. each circle has the same radius.

1. **a. Proportional circles:** the *area* of the circle must represent the total sample size. Choose an appropriate scale for the radius, eg. 1cm = 10 vehicles, fields, people, etc. You must now work out the radius for all the samples.
NB. THE LENGTH OF THE RADIUS FOR EACH CIRCLE IS NOT *DIRECTLY* RELATED TO THE TOTAL SAMPLE SIZE. For example, two samples are to be shown as two pie graphs. There are 20 units in the first sample and 40 in the second. Because of this the *area* of the second circle must be double the area of the first. This doubling of the area is not brought about by doubling the radius.
NB. The area of a circle is πr^2 and not πr.

To work out the appropriate length of the radii:
i. work out the radius using the original scale (1cm = 10 vehicles);
ii. work out the square root of this length. This will be the radius length you require when drawing the circle.
The table below shows the calculation for our particular example.

Sample number	Total in sample	Radius using scale 1cm = 10 vehicles	Square root (Radius length drawn in cms)
1	20	2.0	1.41
2	40	4.0	2.0

As can be seen from this example a doubling in sample size does not give a doubling in the radius length (1.41 and 2.0 cms).

Another alternative to this method is to construct an accurate scale line and this is explained on page 47. In practice the use of a scale line is only worth the effort when there are large numbers of circles to be constructed. This is often the case when they are used on base maps (see pages 48 and 49).

b. Equal sized circles: as already mentioned these are used when sample sizes are identical *or* the component categories are expressed as a percentage of the whole. The use of a simple scale will give the circle size desired and all the circles use the same radius length.

Once the circles have been drawn the following steps are followed for both of the methods described above.
2. Each circle can now be divided into its component categories. A little maths is involved in converting the observed frequencies into angles that can then be used to divide up the circle from the centre. The equation needed uses the same principle as for percentages and is as follows:

Angle of component category =
$$\frac{\text{Observed category size}}{\text{Total sample size}} \times 360$$

The following table shows how the angles have been worked out for the example opposite.

Vehicles passing along Water Street, Chipping, 12th November 1987 12.00 to 12.30 p.m.

Category name	Category size	$\dfrac{\text{Category size}}{\text{total size}} \times 360 =$	Angle on pie graph
Cars	27	$\dfrac{27}{40} \times 360 =$	243°
Lorries	5	$\dfrac{5}{40} \times 360 =$	45°
Vans	3	$\dfrac{3}{40} \times 360 =$	27°
Buses	2	$\dfrac{2}{40} \times 360 =$	18°
Others	3	$\dfrac{3}{40} \times 360 =$	27°
Total size	40	$\dfrac{40}{40} \times 360 =$	360°

Note. An alternative method would be to find the angle corresponding to 1 item (in this case 1 vehicle).

$$\text{Angle for 1 vehicle} = \frac{1}{40} \times 360 = 9°$$

The maths is now very simple, for example 5 lorries = 5 × 9 = 45°

3. Using a protractor divide off the circle by using angles at the centre. The first category starts to the right of 12 o'clock. If there is a "miscellaneous" or "others" category this goes last (immediately anticlockwise of 12 o'clock). If more than one pie graph is drawn then the category order must stay the same.

4. Colour or shade the different sections. If there is more than one pie graph then use the same colour/ shading for each of the component categories. Put on the labels in ink and do not forget the key.

Vehicles passing along Water Street, Chipping, 12th November 1986 12.00 to 12.30 p.m.

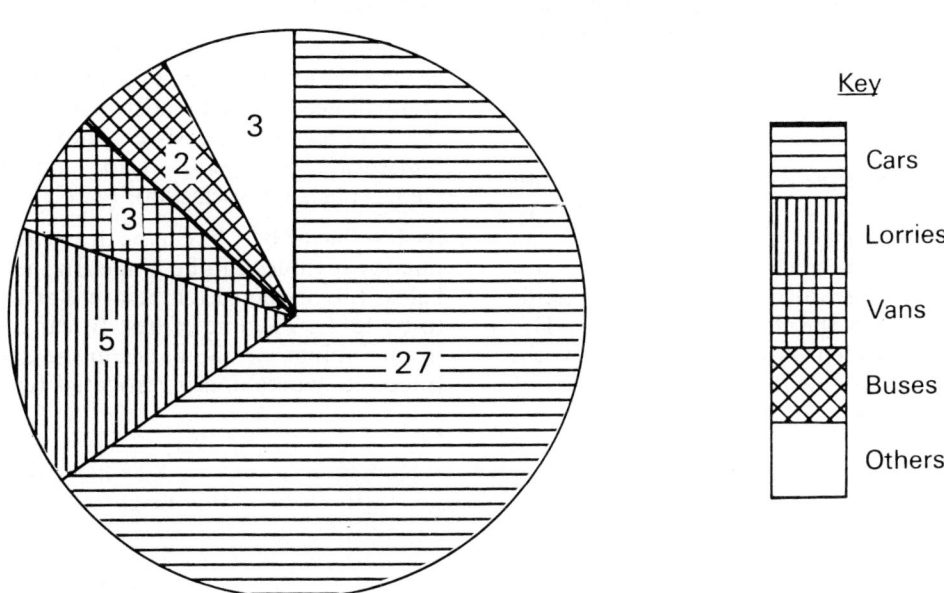

Key

Cars
Lorries
Vans
Buses
Others

Scattergraphs (correlation graphs)

When to use

Up to now most of the graphs described have used single types of observation/measurement, for example pebble size, numbers of vehicles, etc. Scattergraphs use two different types of observation. The different observations are called **variables**. Scattergraphs are used to show graphically any connection/relationship between the two variables. Another name for connection is **correlation** hence the name correlation graph. Scattergraphs are one of the most commonly used types of graph.

Examples

There are literally hundreds of cases where this type of graph might be used as geographical enquiries are often concerned with the connection between paired variables, for example air temperature and altitude, slope angle and soil depth, building age with distance from a CBD.

Method of construction

Here we use the example of altitude against air temperature.

i. Two axes are drawn in the usual way (the scales just exceeding the highest recorded value). Try to decide which variable is causing the change in the other. This is called the **independent variable** and uses the horizontal axis. The variable that is being changed by the other is called the **dependent variable** and uses the vertical axis. In our example, altitude is the independent variable and air temperature is the dependent variable.

ii. Plot each **pair** of figures as a single point using the observed data as co-ordinates. The more points (paired data) the more reliable the graph. DO NOT JOIN UP THESE POINTS.

iii. Study the resulting pattern. The possible patterns are explained in the section Scattergraph Patterns on page 26 and will help you analyse your graph. You may be able to put in a 'Line of best fit' using a sharp pencil and ruler. This is done by trying to draw in by eye a straight line that appears to pass as close as possible to all the points plotted. This is not possible if your graph shows little or no correlation. The line does not have to pass through the point of origin (where the two axes meet).

Note. The line of best fit can be put onto the graph accurately using a statistical test called regression but you will need a good statistics manual that shows you the process (see the sister booklet to this manual, *Statistical Analysis*, also published by the Geographical Association).

iv. If a line of best fit can be drawn then predictions of unknown values can be made. In the worked example below the predicted temperature at 250 metres is seen to be 15°C. The line of best fit can also highlight any 'rogue points' or anomalies in the data. These can be traced to source and investigated further.

Worked example

What is the connection between altitude and air temperature?

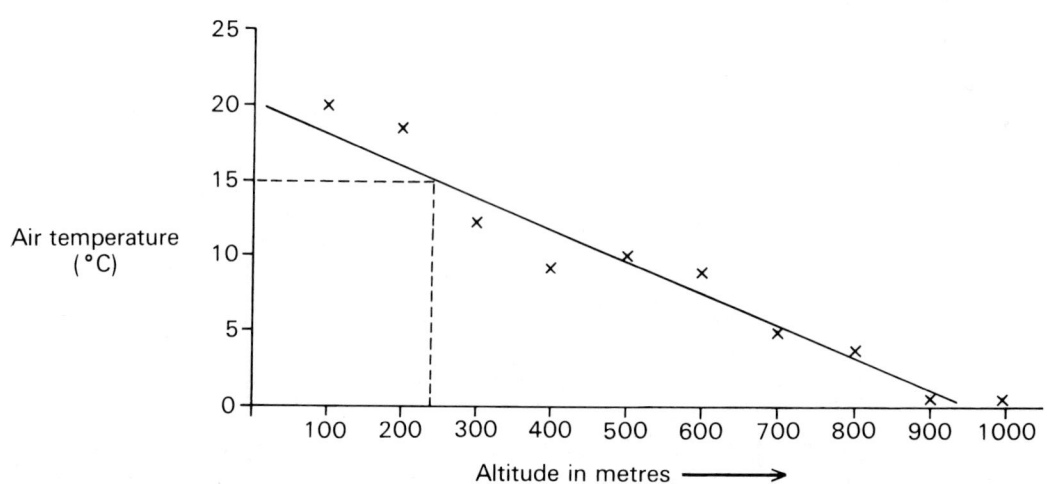

A Scattergraph Variation

When to use

The scattergraph method already described is used to show a relationship or connection between two variables that have been expressed quantitatively. It is equally possible to use a similar method to show the connection between two variables where only one of the variables has a quantitative value, the second variable having a non quantitative value (land use, geology, soil type, etc.).

Examples

Any two connected variables where one has a non quantitative value,

Land use		Altitude
Vegetation type		Gradient
Rock type	**against**	Soil variables
Soil type		(soil pH etc)
Building materials		Distance
Building functions	**against**	from
Environmental factors		CBD

Worked example

Method of construction

Here we use the example of land use against altitude.

i. Two axes are drawn in the usual way.

ii. The quantitative variable (altitude) usually uses the vertical axis and is labelled as any other numerical scale.

iii. The non quantitative variable (land use type) uses the horizontal scale. Each of the different land use types is allocated a 'block' of equal width. The order across the scale may be important, for example if there is a preconceived idea (hypothesis) as to the expected connection between land use and altitude then the categories may be placed in the anticipated order.

iv. The observed values are now plotted in a similar manner to scattergraphs (see the worked example below).

v. It is not usual to fit a line of best fit, rather 'zones' and 'overall trends' are identified, for example the altitude zones of each land use type, its upper and lower limits, merging boundaries and patterns of land use with an increase in height.

Is there any correlation between altitude and land use?

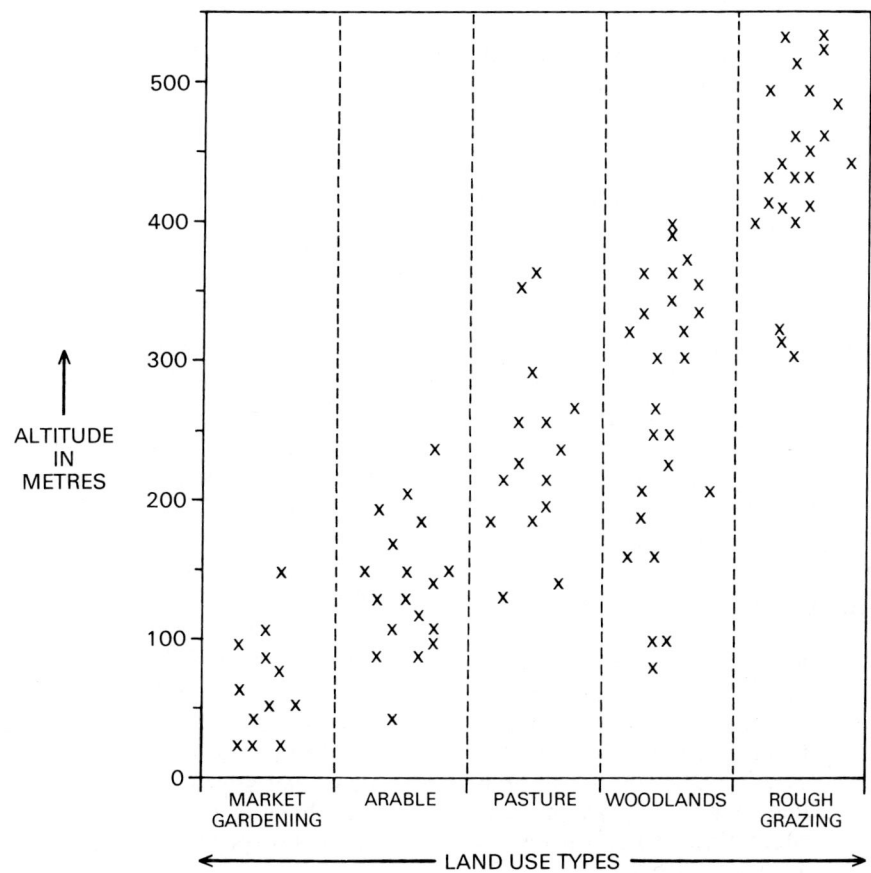

Note. All types of scattergraphs at first sight appear to be very useful when it comes to hypothesis testing, especially where correlations are involved, but they must be treated with care. Read the next section carefully.

Scattergraph Patterns

Some basic rules can be applied. The closer the points are to a single straight line the better the correlation (connection) between the two variables. The following graphs show what some patterns may mean:

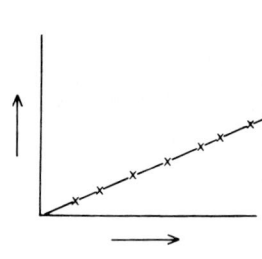

Perfect correlation. All points lie on the line of best fit. In this case the correlation is a positive one but it equally applies to negative correlations. (A correlation of either + 1.0 or − 1.0). You cannot get a better connection between any two variables than this.

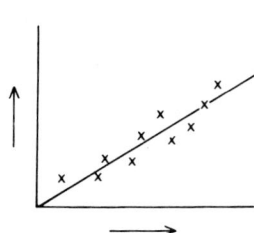

Positive correlation. Not a perfect correlation but the points are close to the line of best fit. This is a positive correlation, as one of the variables increases in size so does the other, eg. wind speed and altitude, stream order, discharge, cross sectional area with distance downstream, etc.

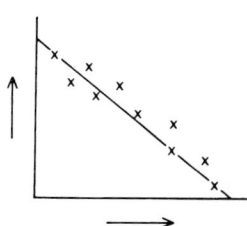

Negative correlation. Similar to the previous graph except as one variable increases in size the other decreases in size, eg. air temperature and altitude, stone size with distance down a stream, etc.

These three example graphs show some form of correlation between the two sets of variables. Not all the graphs you draw will show this so here are two more examples to show what a graph looks like when there is little or no correlation.

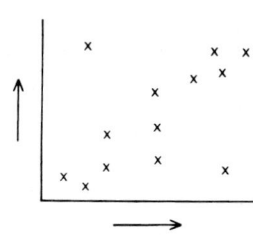

Poor correlation. There is some sort of connection but the points are some distance from a straight line. In this case it is difficult to put on a line of best fit. Graphs that show these sorts of results are inconclusive.

More data to be collected.

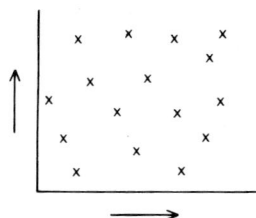

No correlation. The points appear to be at random, there is no real straight line pattern. In this case the results show that there is no connection at all between the two variables.

Warning: Scattergraphs can have limitations

i. It is often difficult to put on a line of best fit by eye. The regression analysis is much more accurate but can be complex.

ii. Not all correlations between variables are straight lines (linear). Some appear as curves which makes the fitting of a regression line difficult. The example below illustrates such a relationship.

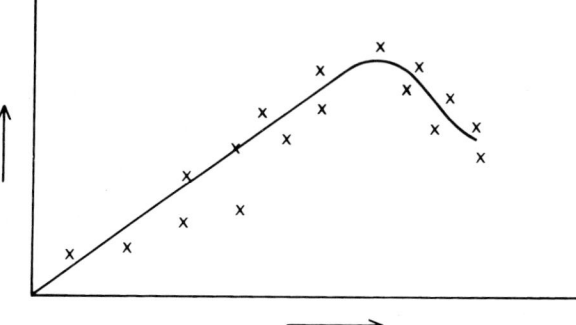

The answer to this problem may be to plot the results either using the logarithmic values of the data or by graphing on logarithmic paper (see pages 34 and 35).

iii. What might appear to be a strong correlation does not always imply that one variable has influenced the other because:

a. the resulting graph may have been caused by chance. By experimenting with various pairs of data it is often possible to get a 'nonsense conclusion'. Statistics can be misleading with two sets of variables giving a good correlation even though it is obvious that the two variables could not ever be connected. For example it can be shown that the birth rate in Sweden changes directly with the number of stork nesting sites!

b. the correlation is caused by a **third variable** that has not been considered, for example, the temperature drop that appears to be caused by a rise in altitude may have instead been caused by a third variable, i.e. wind speed that is also related to altitude.

USE AND ANALYSE THESE GRAPHS WITH EXTREME CARE AS THEY CAN BE MISLEADING

More worked examples

Each of the following graphs shows a different type of correlation. What type is each, how close are the connections and can we explain the patterns? Note which variables are independent and which dependent.

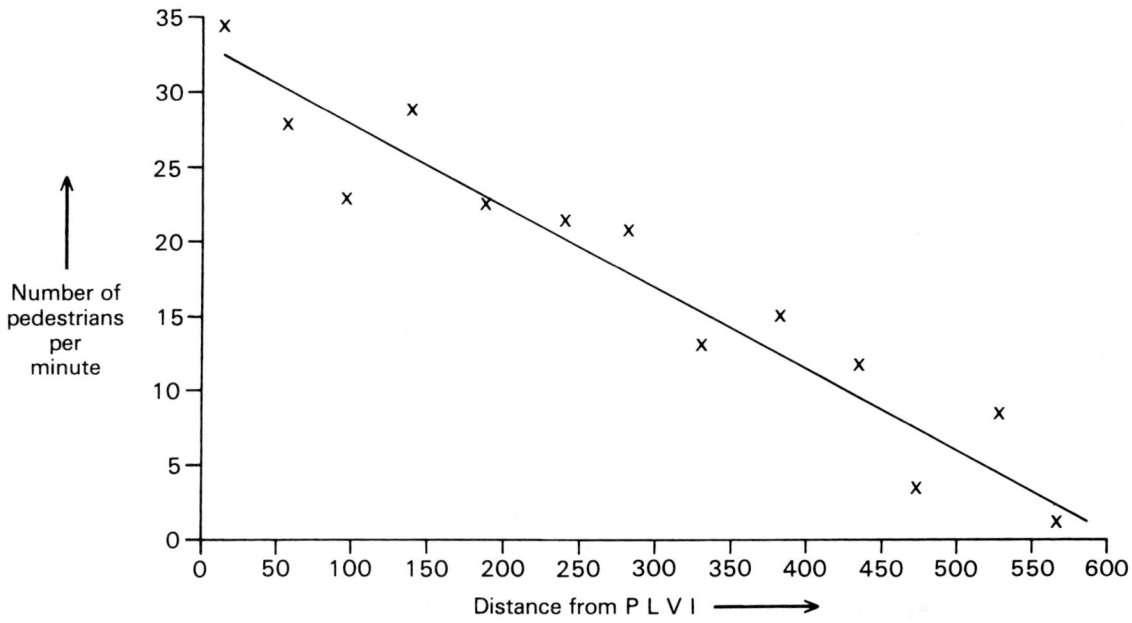

Pedestrian densities with distance from the peak land value site of an urban CBD

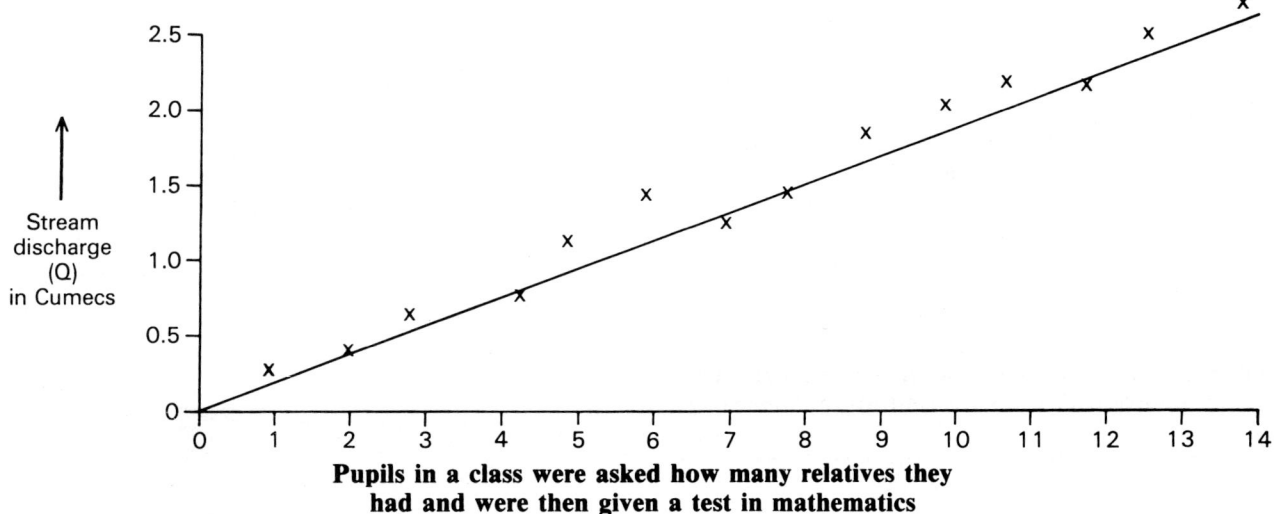

River discharge with distance downstream from the source

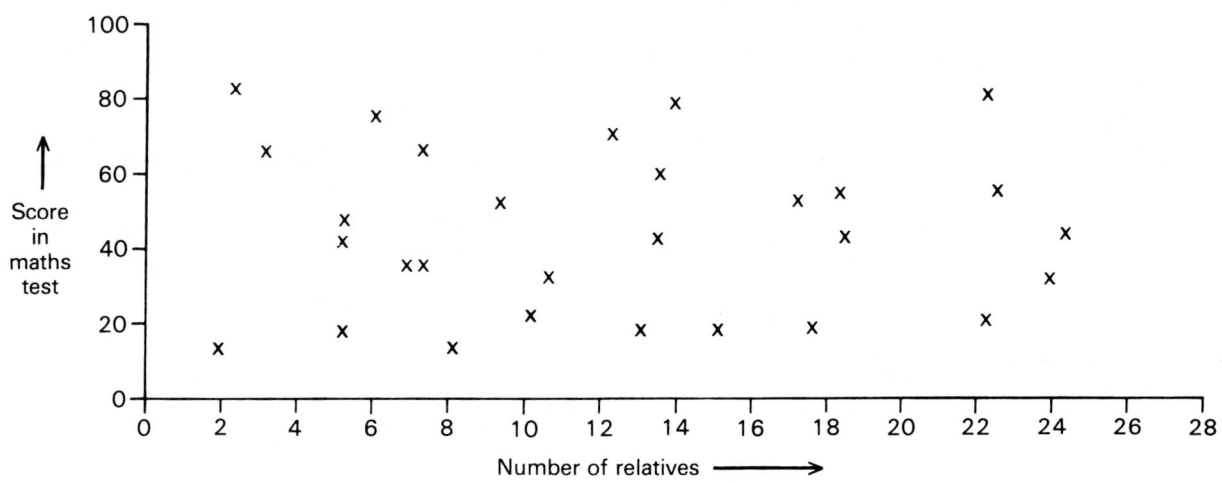

Pupils in a class were asked how many relatives they had and were then given a test in mathematics

Displaying Orientated Data — Rose Diagrams

When to use

Rose diagrams are sometimes called 'Star diagrams' and are used when the observed data takes the form of a compass direction or bearing. The end result is a visual impression of any predominant orientation that the data may display. The significance of this orientation can then be tested statistically using either a Vector analysis or a chi squared test.

Examples

Wind roses to display predominant wind directions and strengths, sediment analysis (orientation of fluvial, marine, glacial, talus deposits), the morphometric analysis of such features as drumlins, corries, shakeholes, etc.

Method of construction

There are several different ways of constructing rose diagrams depending on the nature and complexity of the data involved. In nearly all cases the data has to be put into pre-arranged categories, for example north, north east, east, etc. *or* $0-9°$, $10-19°$, $20-29°$, etc.

The most commonly used methods are shown below along with worked examples.

1. Simple roses.

The compass direction is taken either over a given time or for a particular sample size. The simplest type of rose puts these directions into categories that represent the eight points of the compass.

A scale is worked out and the length of the arm represents frequency. All the arms are drawn the same width.

Worked example

Wind directions observed at Hothersall Lodge Field study Centre over a time period of one month

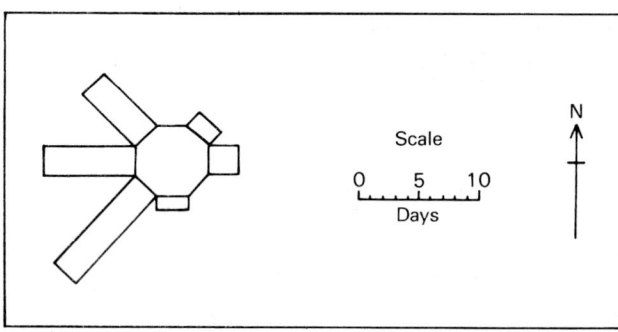

It is possible to be more precise by using the 16 points of the compass or even size classes ($0-9°$, $10-19°$, $20-29°$ etc). This method is discussed in greater detail later on in this section.

2. Compound roses.

Sometimes it is possible to display *several* variables on one graph, for example wind direction, frequency and strength.

The *length* of the bar represents the frequency, and now the *width* of the bar is varied to show wind speeds/strengths. In this particular case the number of categories available is restricted by the second variable (wind speed) being included.

Worked example

Wind direction and strengths observed at Hothersall Lodge Field study Centre for 1984

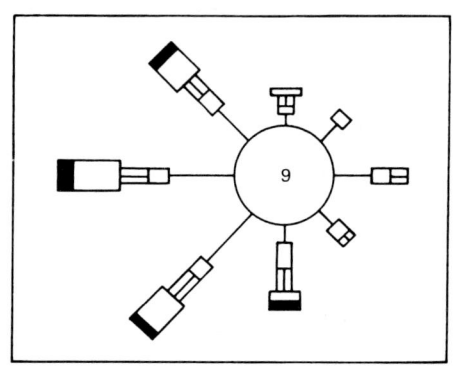

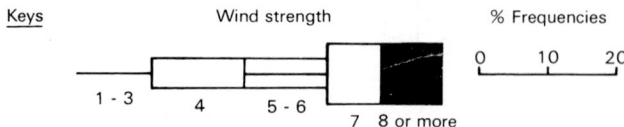

Legend
i. The length of each line represents the percentage frequency of each wind direction.
ii. Wind strengths are from the Beaufort scale and are shown in the key below.
iii. The number in the circle represents the percentage frequency of calms.

3. Azimuths

An azimuth is a compass bearing and azimuths are most commonly used when sediment or morphological orientations are displayed. The data is collected in 'raw' form in the field as an exact compass bearing.

The raw data is now categorised into classes, for example 0 – 9°, 10 – 19°). These classes are called **azimuthal classes**. A table showing azimuthal class and frequency is useful at this stage. The class interval chosen is important. If it is too large (eg. every 40° or 50°) then patterns appear that are not a true reflection of reality, but if they are too small then no patterns (preferred orientation) may be shown even if they actually exist in reality. By the same argument sample size is equally important, the larger the sample the better.

Once the data has been categorised then the graphs can be drawn.

i. A straight line is plotted along the direction of the **mid point** of each azimuthal class, for example the mid point of the azimuthal class 20 – 29° is 24.5. The length of the line is therefore related to the frequency of each class.

ii. These lines may be left like this as in diagram 1 below; the ends of each line can be connected together and shaded in as in diagram 2; or each azimuthal class is shaded in up to the end of the line as in diagram 3.

In practice it is the third method that is most commonly adopted. The concentric circles are included to give an idea of frequency.

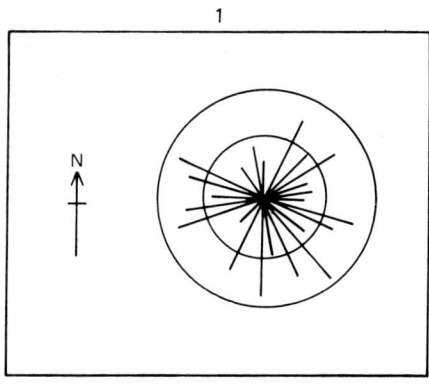
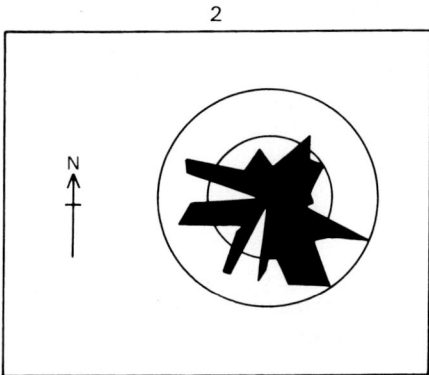
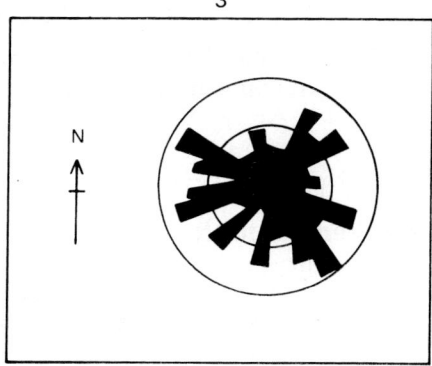

1   2   3

Reflected, half and single roses

With certain data each feature may have *two* opposite bearings, for example the long axis of a drumlin may be 10° and 190°. Other examples are the axes of sediments, shakeholes, roads and streets, etc. The resulting rose would therefore show two reflected/mirrored halves. Many fieldworkers therefore only construct half the rose (0 – 180°) when this data is

involved. Examples of this are shown below.

Other sets of data have only one possible bearing eg. wind direction, corrie orientation. In these particular cases a full rose (0 – 359°) is constructed. An example of this is also shown below.

Finally it must be stressed that these diagrams are only valid if a large sample is involved.

Worked examples

1. Long axis orientation of bedload in an upland stream — arrow indicates the direction of water current.

2. Wind rose for 1962 – 1966

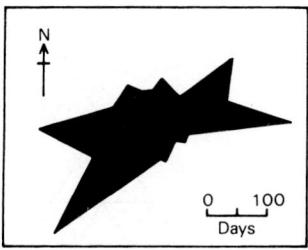

1. Reflected   2. Half rose (using same data)   3. Single rose

Miscellaneous Graphs

There are some types of graph that have a limited application but are well worth a mention at this point.

1. Triangular graphs

These are sometimes called **ternary diagrams** and have three axes instead of two. They take the form of an equilateral triangle. The important features are:

i. Each axis is divided into 100 representing percentages.

ii. From each axis lines are drawn at an angle of 60° to carry the values across the graph.

iii. The data must be in the form of three components, each component representing a percentage value and the three percentage values adding up to 100%.

Their main value arises when data for several locations is plotted on one graph, the relative position of the points giving a quick visual impression of the relative dominance of one component or another. Care must be taken when plotting and interpreting such a graph for it can be confusing at first acquaintance.

Example

Employment structure (primary, secondary and tertiary), soil particle size (silt, clay and sand), agricultural land use (arable, pastoral, others) etc.

Worked example

Diagram of a triangular graph showing the way in which the values are carried across the graph

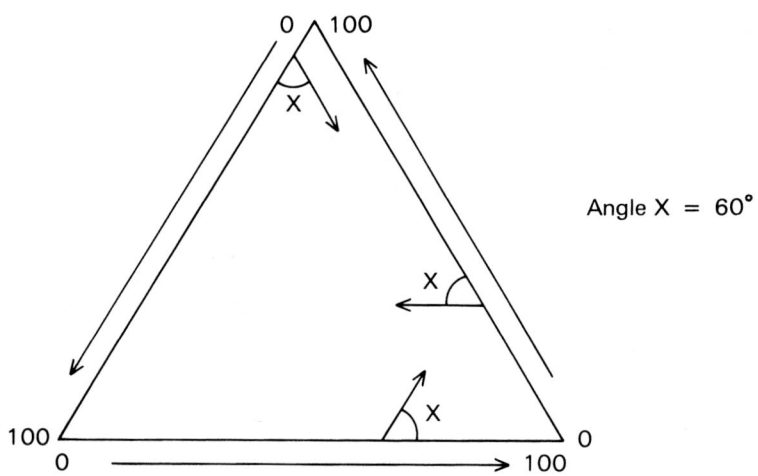

Angle X = 60°

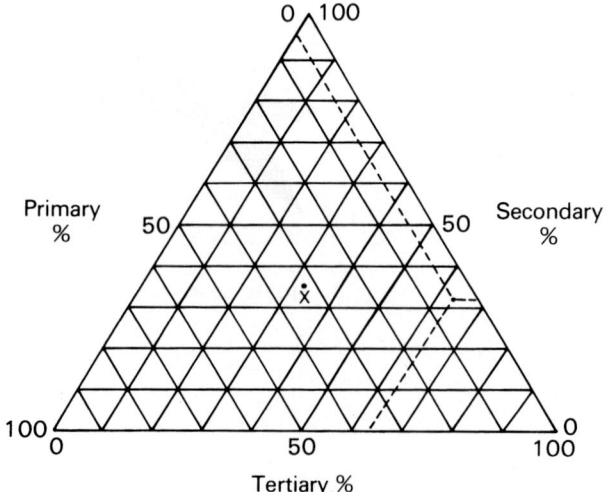

Triangular graph to show employment structure of a county. (Figures below)

PRIMARY	5%
SECONDARY	33%
TERTIARY	62%

Point X marks the spot where all three elements are identical

2. Circular graphs

These are sometimes used to show a variable that is continuous over time, such as temperature data. On a normal line graph there is a false break. It starts at one end on January 1st and ends at the other end on December 31st. On a circular graph this break does not occur.

Circular graphs are easily drawn. There are two axes — the circumference of the circle usually represents time, such as months of the year and the radius which represents the observed values, eg. temperature, oxygen content of a pond, etc.

The main drawback to this method is that change over time is shown by the line's relative position to the centre of the circle and the interpretation at first can be more difficult than the straightforward rises and falls of a line graph. Familiarity will however increase competence in interpretation.

Examples

Yearly temperature and air pressure readings, yearly variables of a pond, traffic flows, any continuous variable over a time period.

Worked examples

Average monthly temperatures for Moscow, Russia

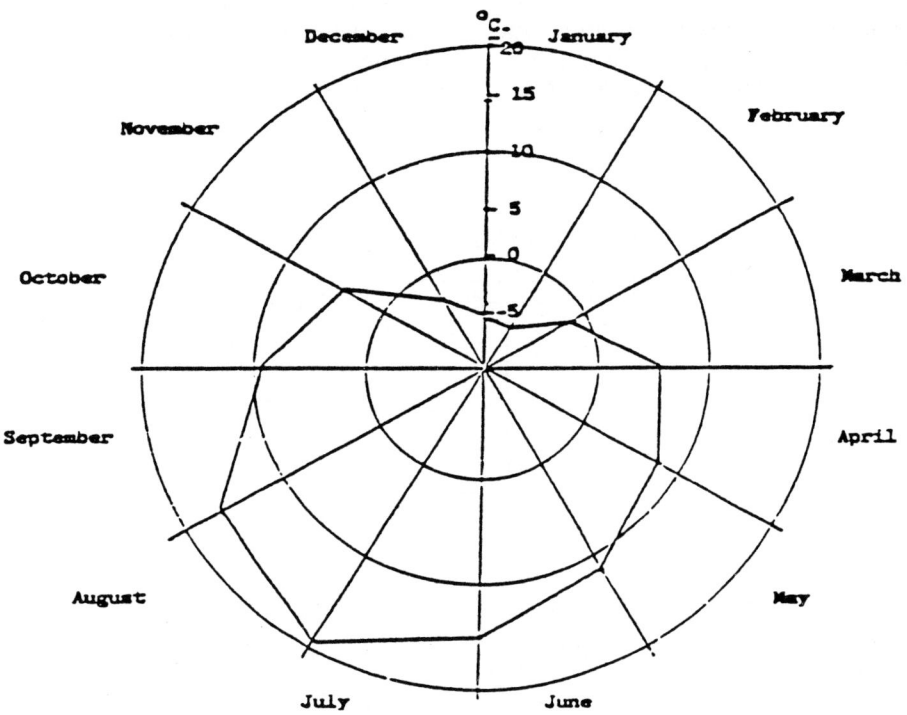

Average monthly temperatures and rainfall for Lusaka, Zambia
Source: L.R. Frost, *Teaching Geography*, June 1985

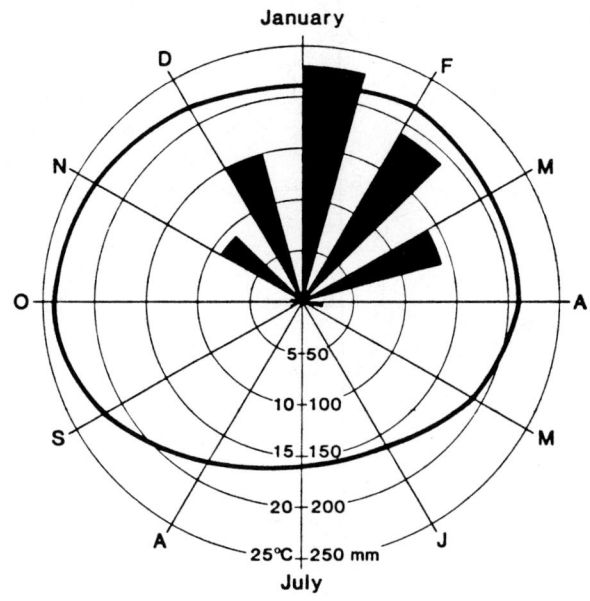

31

3. Polar co-ordinates

This type of graph is related to both rose diagrams and scattergraphs, both of which have already been described in earlier sections. In many cases it is actually referred to as the scattergraph equivalent of the rose diagram.

In appearance it is rather similar to the circular graph already described on page 31 but there are some very important differences.

   i. As in the case of scattergraphs there are always two variables of which one is always **orientation**. This is expressed as a compass bearing from north (as in the case of the rose diagrams).

  ii. The second variable is represented by distance from the centre of the graph and can represent any variable quantity, for example size, altitude, etc.

 iii. A third dimension may be introduced *either* by varying the size of the symbol used (dot, cross etc) to show abundance *or* by the use of different symbols for different categories of observation (see the worked example).

Examples

It is a very specialised type of graph and can only be used where the data shows some form of orientation as one of its variables. Examples may include: *corries* — aspect against altitude, aspect against size; *vegetation types* — aspect of slope against gradient against frequency of occurrence; *slopes* — aspect against any soil variable (ph, water, humus content, etc.).

Worked example

The example below shows orientation of corries in North Wales against their altitude. The corries have been divided up into the three main mountain ranges included in the survey. These are shown by using three different symbols.

Corrie orientation against altitude in North Wales

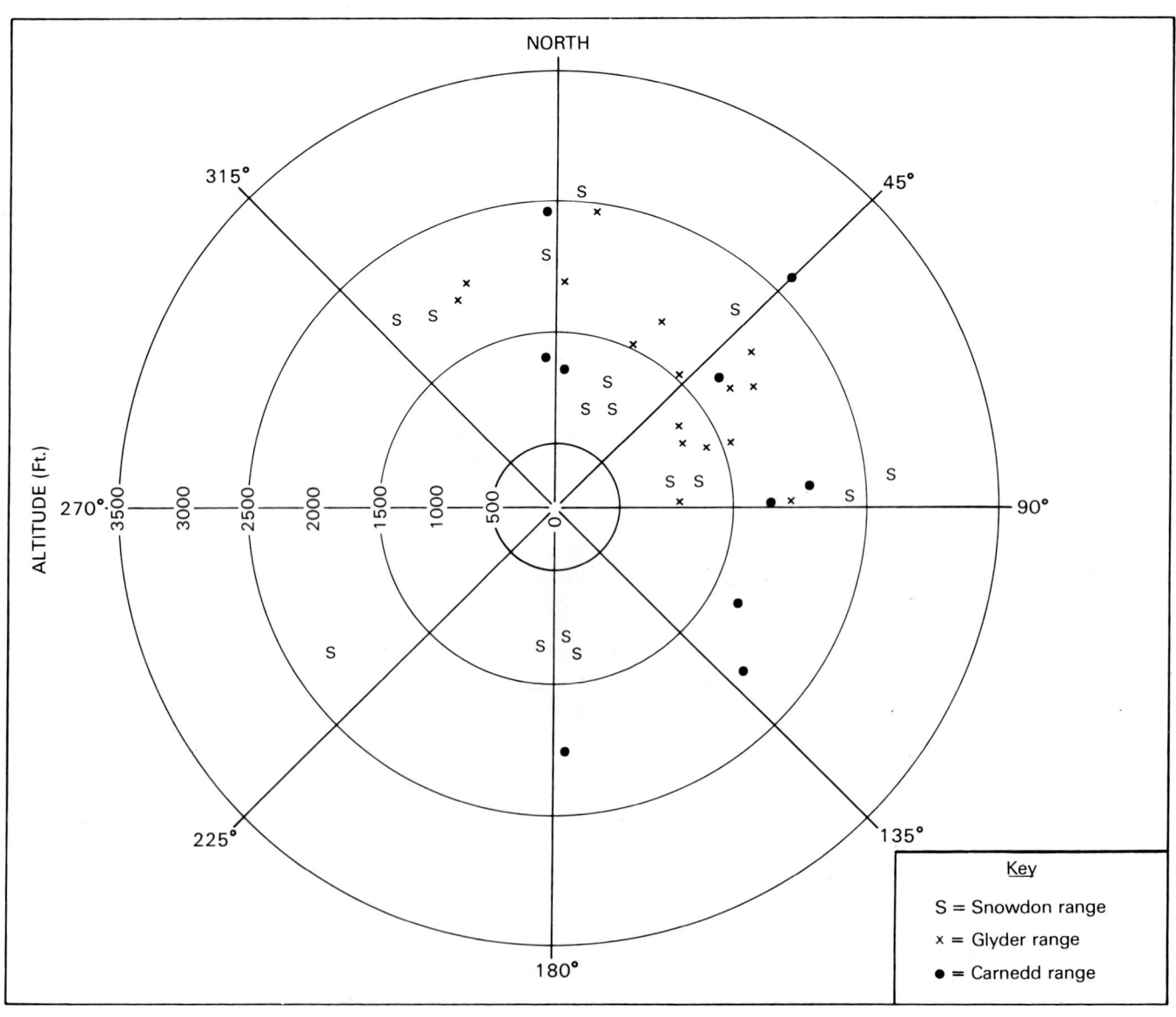

Key

S = Snowdon range

x = Glyder range

● = Carnedd range

32

4. Reverse bars

These particular graphs are related in appearance and method of construction to barcharts and histograms. They are commonly used when negative values are represented as well as positive values. As will be seen below these graphs are especially useful when we want to show trends/changes over time and/ or space.

The method of construction is fairly straight-forward. The horizontal axis represents the categories involved whilst the vertical axis is drawn above *and* below the end of this horizontal axis. Values ranging above the horizontal are positive whilst those ranging below the horizontal are negative values.

As can be seen from the second of the worked examples this particular type of graph can be shown quite effectively on base maps.

Worked examples

Population changes in Lancashire's urban and rural areas, 1971 – 1981

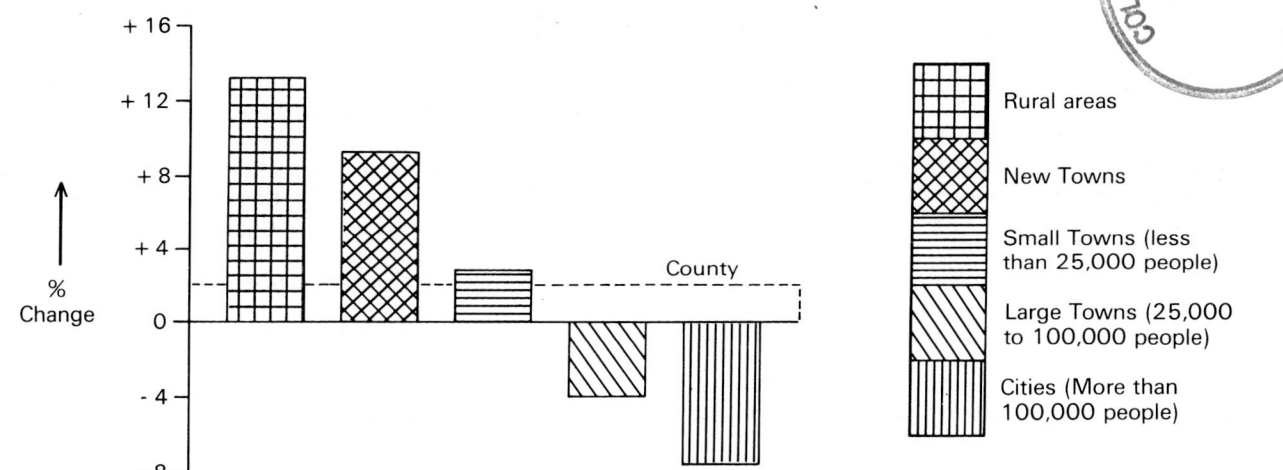

Rural areas

New Towns

Small Towns (less than 25,000 people)

Large Towns (25,000 to 100,000 people)

Cities (More than 100,000 people)

Population changes in selected age groups, Lancashire districts, 1971 – 81

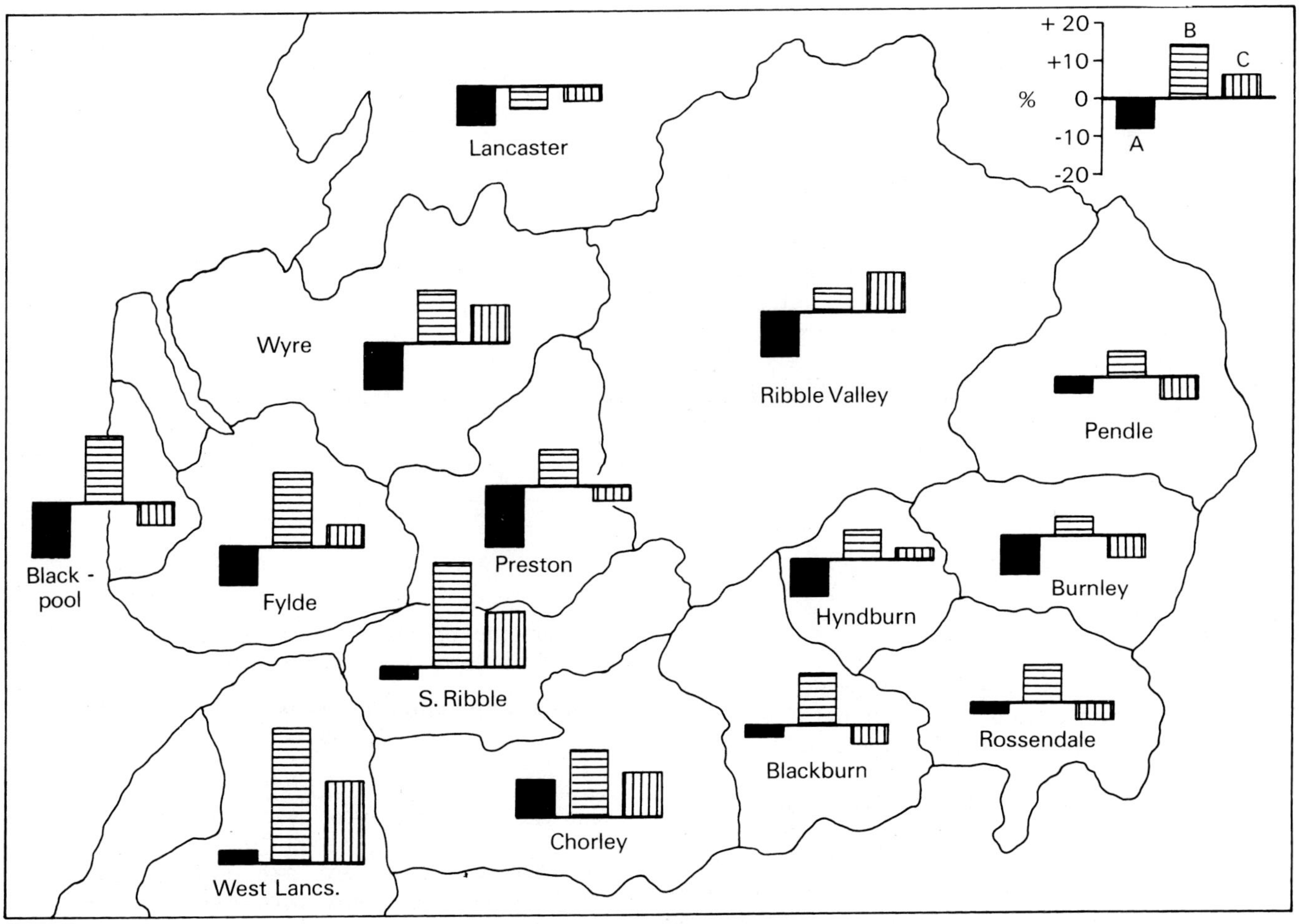

A = 0 - 15 (Preschool / School)   B = 16 - 24 (Young adults)   C = 60 / 65+ (Retirement)

33

Logarithmic Graphs

Logarithmic graphs can provide a useful tool when displaying and analysing some fieldwork data. The values may be plotted directly onto logarithmic graph paper or by plotting the logarithmic values of the raw data onto normal (arithmetic) graph paper. Both methods will be illustrated later in this section.

When to use

There are several valid applications of this technique although for our purposes there are two main situations in which the use of logarithmic graphs is helpful.

i. When the range of values of one or both of the scales is so large that a meaningful graph could not be plotted onto a sheet of linear scale graph paper.

ii. Where the purpose of the data collection is to compare *rates* of growth of variables.

Examples of use of logarithmic graphs

Representation of economic data, changes in population size, agricultural yields, energy production, graphs where the rates of increase or decrease are compared.

Method of construction

As mentioned in the introductory paragraph specially printed logarithmic paper may be used (this should be relatively easily available). Alternatively the logarithmic values of the data may be plotted onto arithmetic graph paper (with a pocket calculator this technique is simple and straightforward).

Use of logarithmic paper

There are essentially two types of logarithmic paper. Fig. 1 shows paper with one logarithmic and one linear scale, (semi-log or log/linear paper). Fig. 2 shows paper with two logarithmic scales, (log/log paper).

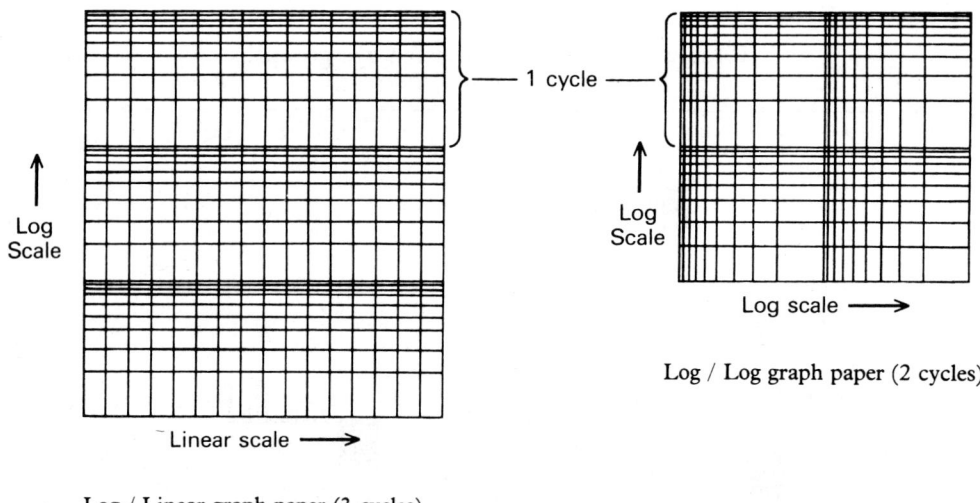

Log / Linear graph paper (3 cycles)

Log / Log graph paper (2 cycles)

Fig. 1.

Fig. 2.

Examples of use

The following values for population growth of three towns were obtained from census material. The aim of the data collection is to compare the growth rate of each town.

It would be an advantage to display the population of each town on the same graph so that visual comparisons are possible but the values (from 3,000 to 158,000) are dispersed over too wide a range for a meaningful linear graph to be constructed. Either the graph paper would have to be so large as to be

cumbersome or the scale so reduced that the line plotted for town X would be quite inaccurate. In this

Table 1. Population figures for three towns, X, Y and Z, 1900 – 1980

Town	1900	1910	1920	1930	1940	1950	1960	1970	1980
X	3000	4000	5200	6800	8900	11500	15000	20000	26000
Y	8100	10500	18500	24000	24000	31500	41000	54000	70000
Z	80000	87000	94000	104000	112000	121000	133000	145000	158000

case log/linear paper is a useful tool.

The first decision is to set a base line. A logarithmic graph does not include zero. With each diminishing cycle the values decrease but never actually reach nought. Convenient base lines are to the power of ten, eg. 0.1, 1, 10, 100, 1,000, 10,000, etc. Choose one immediately below the lowest value in your table. In Table 1 the lowest value is 3,000, hence the most convenient base line is 1,000.

Next the scales must be inserted. In our example the x axis is a conventional linear scale and is used to represent time (years). The y axis is divided into a series of cycles or decades, each cycle representing a multiple of ten. Therefore in our example with a base line of 1,000 the next cycle begins at 10,000 and the third at 100,000. For our range of values three cycle paper is adequate although for greater ranges paper with more cycles is available. Six cycle paper would allow a range of values from one to one million.

Finally the values are plotted directly onto the paper.

There is one further advantage of the logarithmic graph in addition to allowing us to plot a wide range of figures. The graphs actually show us the rate of change and so here it can be seen from the inclination of each plotted line that towns X and Y have identical growth rates which is higher than town Z. This fact may not be immediately obvious from examining the data alone.

In certain cases it will be advantageous to use graph paper with two logarithmic scales, eg. when both variables exhibit a wide range of values or when confidence limit lines are required on a divergent scatter graph. The latter is outside the scope of this book and reference to more advanced statistical books is advised.

The construction of a graph with two logarithmic scales is the same in principle as the construction of the logarithmic scale of the worked example.

Log/linear graph showing the population increases of three towns, 1900 − 80

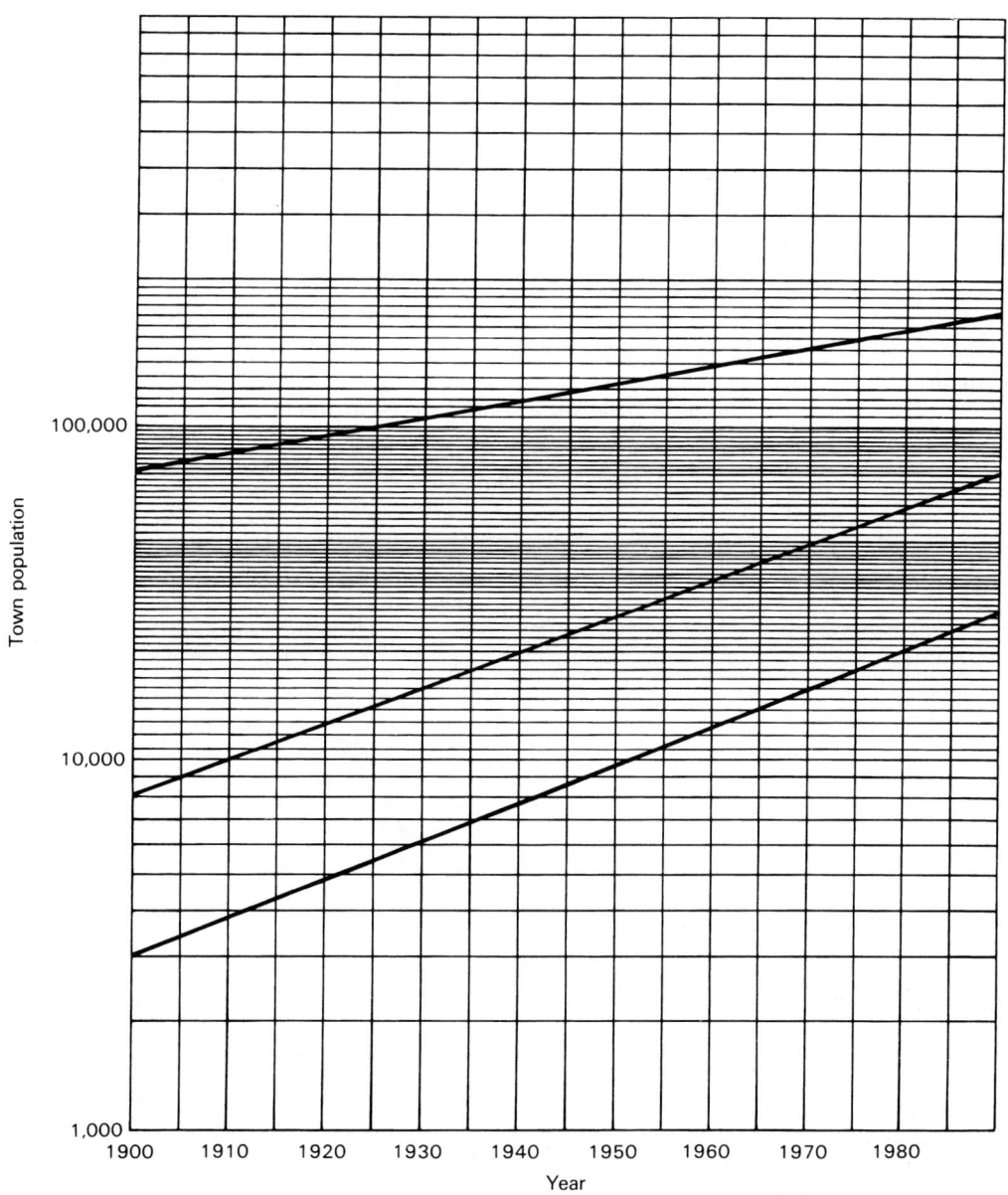

Profiles and Cross Sections

When to use

It is a very important skill to be able to construct accurate **side views** of particular geographical features such as river channels and beach profiles. Several can be combined to show change in shape and size from one location to the next.

They are often used as base diagrams for displaying other related data, for example stream velocities across a stretch of meander, pebble sizes and shapes along a beach. Cross sections are frequently drawn in order to calculate accurate areas. By combining these with other measurements volumes can also be calculated, for example grike sizes on a limestone pavement.

Examples

Profiles — slopes, valleys, beaches, corries, waterfalls, etc.
Cross sections — river channels, gullies, limestone pavements.

How to construct

Profiles and cross sections appear to be similar but there are important differences and they are therefore dealt with separately.

1. Profiles — data has been collected for these usually by two different methods.

i. By dividing the feature into separate, definite, straight sections and then taking measurements of gradient and length for each section.

ii By taking an accurate gradient measurement at regular intervals, for example every metre. This is usually carried out either by using a pantometer or by accurate surveying techniques.

To construct.
i. Work out the total length of the profile to be drawn. Work out a suitable scale so that the profile once drawn fits neatly onto the graph paper. Remember because of the angles the total length will be 'dragged in' once drawn.
NB. If more than one profile is to be drawn use the same scale throughout.

ii. Start at one end of the profile. Draw in the first set of measurements using a protractor for the angle and a ruler and scale for the length.

iii. Work your way along the profile in the same manner until completed.

Worked example

Slope profile on a limestone escarpment

Scale : 2 millimetres = 1 metre

LENGTHS (m)	7	16	10	8	11	21
ANGLES (°)		5	12	65	4	-3

SECTION	1	2	3	4	5	6
pH	6	5	8	8	7	6
% WATER	64	75	45	23	52	49
% HUMUS	7	5	1	1	4	6
DEPTH	35	41	2	2	29	31

The profile in this particular example shows the shape of the slope whilst the table below it records some of the variables measured in each section. Scattergrams can now be constructed to try and see if any connections/correlations occur between any of these variables.

Cross sections — these are usually constructed by measuring the width of the feature and then taking depth/height measurements at regular intervals across this width. The greater the number of measurements the more accurate the finished product.

To construct

In this particular case two scales are needed (the vertical and the horizontal). If possible they should be identical but sometimes the shape of the feature is rather shallow (eg. a small stream) in which case it can only be seen in cross section if the vertical scale is exaggerated. This may cause problems if the resulting cross section is needed for further calculations, eg. wetted perimeter, hydraulic radius, etc. This problem is discussed in greater detail on the next page.

i. Take the width of the feature and work out a scale which will allow the width to fit onto the graph paper. This is the horizontal scale.

NB. If more than one section is to be drawn for comparison make sure that the horizontal scale and the vertical scale stay the same throughout. Draw a line across the page to represent the width.

ii. On the width line mark off the interval points at which the height/depth measurements were taken.

iii. Work out a suitable vertical scale. Does it have to be exaggerated and if so by how much? Construct a scale line to the left of the horizontal width line and label depth/height measurements according to your vertical scale.

iv. Using the width interval points and this vertical scale line mark off the depth/height measurements as small, neat pencil crosses.

v. Join up these crosses by hand, not with a ruler as in real life these sections are rarely straight, regular lines.

Worked example

Cross section of a small stream channel

Depth measurements were taken at 1 metre intervals across the stream.

1. Horizontal and vertical scale identical — 1 metre = 1 centimetre.

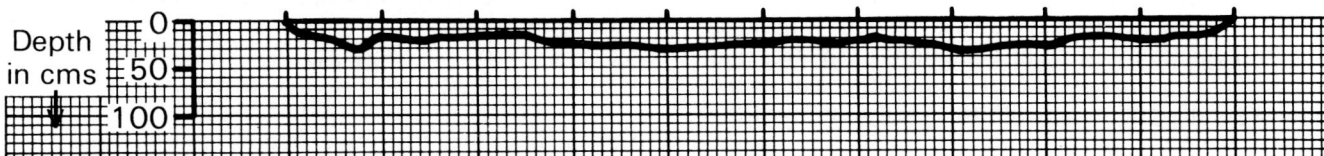

2. Vertical scale exaggerated by 5 — ie 1 metre = 5 centimetres. (Horizontal scale as in first example).

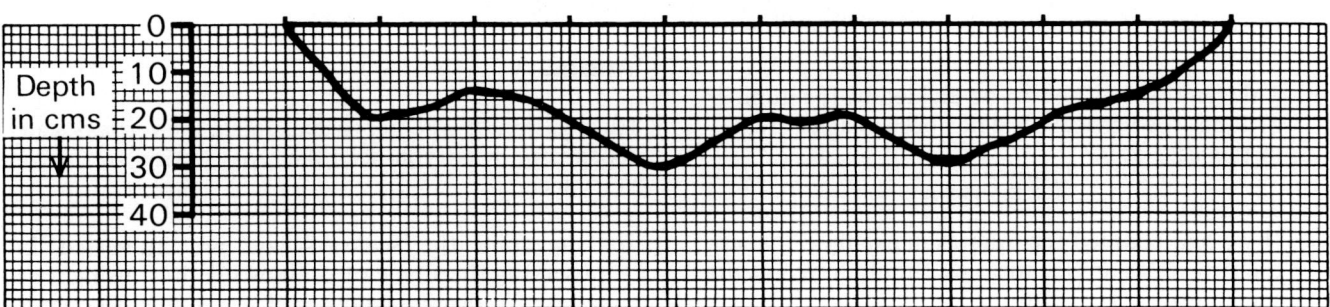

Using cross sections for further calculations

It is possible to use accurately constructed cross sections to work out other characteristics of the feature. Some of these calculations can be made with an exaggerated vertical scale whilst others cannot.

Cross sectional area (CSA)

This is quite an important calculation as it is often used as an indicator of size and is an integral part of the calculation of river discharges (Q).

The CSA can be calculated by counting the squares on the graph paper that appear inside the actual cross section. If we look at the two examples on the previous page we see that in example 1 where the two scales are identical there are less squares to count *but* it is much more accurate to use example 2 where the vertical scale has in this case been exaggerated by 5. This is because the error caused by estimating fractions of squares that are not fully inside the channel (ie. on either side of the line) is greatly minimised with this exaggeration of the vertical scale.

Method

i. Count up the number of squares on the graph paper.
   In the case of example 2 there are 938 millimetre squares.

ii. This number has now to be divided by the number of times the vertical scale has been exaggerated — in this case 5, so
   $$\frac{938}{5} = 187.6$$
   If we had counted the squares in example 1 where the two scales are identical we should have got the same answer, ie. 187.6.

iii. The scale 1 centimetre = 1 metre means that each millimetre square on the graph represents 0.01 square metres in reality. Multiply the number of squares (187.6) by this 0.01 to get the cross sectional area in square metres.
   $$187.6 \times 0.01 = 1.876 \text{ square metres}$$

Wetted perimeter (WP)
In the previous calculation exaggerating the vertical was not a problem as at some stage the total is divided by the degree of exaggeration. If however we want to calculate the wetted perimeter (ie. that part of the channel that is in contact with the stream, bed and bank) we *must* use the *same scale* for the vertical and the horizontal axes.

Method

i. Use either a fine piece of string or a pair of dividers to measure the length of the channel that is in contact with the water. This was done from example 1 on the previous page.
   Wetted perimeter on cross section = 10.2 cms
   Wetted perimeter in reality = 10.2 metres.

Hydraulic radius (HR)
We can now use the cross sectional area and the wetted perimeter to calculate the efficiency of the channel. This is often called the hydraulic radius and the equation is

$$\text{Hydraulic radius} = \frac{\text{Cross sectional area}}{\text{Wetted perimeter}}$$

We can now use the figures already calculated to work out the efficiency of the channel used in our previous worked example.

$$\text{HR} = \frac{01.876}{10.200} = 0.164$$

The higher the hydraulic radius the greater the efficiency of the channel to minimise friction between the channel and the stream flow. This particular example has a poor efficiency which is what we would expect from such a shape.

Summary

The decision of whether to exaggerate the vertical scale depends on what you need the cross section for in the first place.

i. Exaggerate the scale if you are interested in observing the shape of shallow features, using it as a base for displaying other observed measurements or for accurate calculations of the cross sectional areas of a small/shallow stream.

ii. Use identical scales if the feature is large or if you want to calculate the wetted perimeter or the hydraulic radius. Many people draw both types for each set of data.

Using profiles/cross sections as a base for displaying other results
This is particularly useful when **transects** of one form or another are carried out with measurements being taken at regular intervals across the feature (see section on Transects). Below are examples of how profiles and cross sections can be used to display this type of material.

Worked example

A transect to show vegetation change up a slope

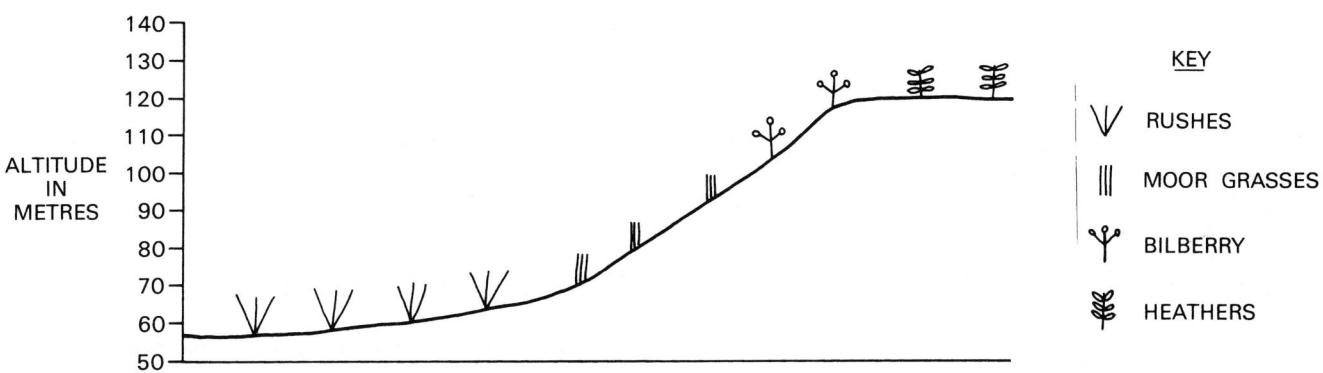

Worked example

At three sites across a meander 100 pebbles were sampled and allocated to size classes. Histograms were constructed on the meander's cross section at the exact points of sampling. Size classes increase from left to right.

See also the section on isopleths (page 52).

Stone size distribution across a river meander

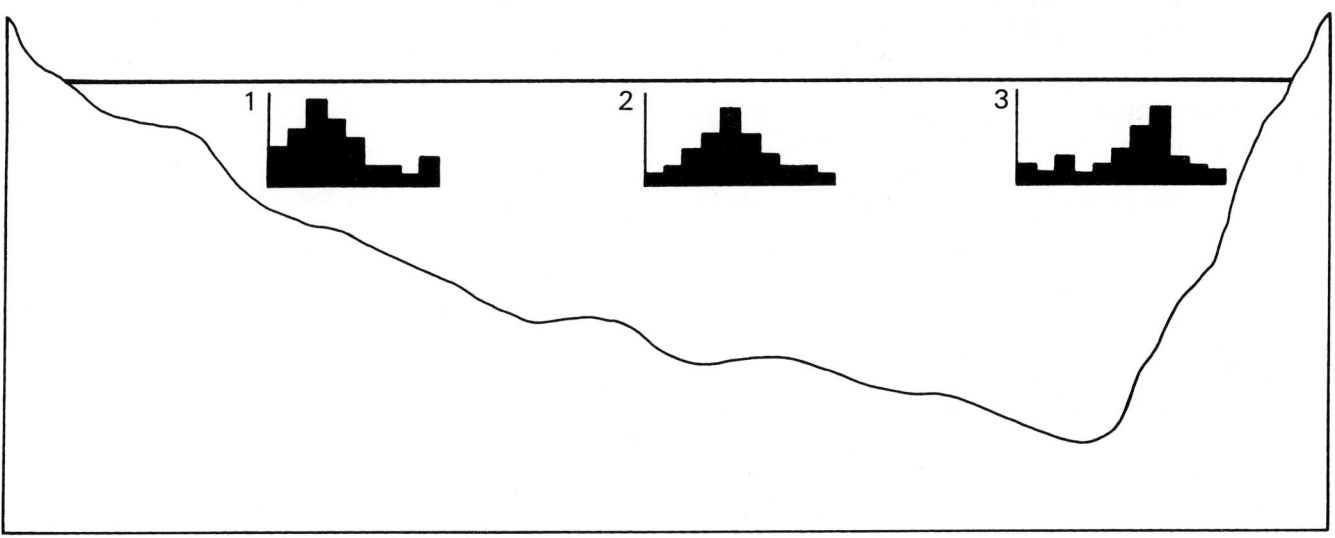

Transects

Transects are commonly used in fieldwork and as such deserve a section to themselves. There are several ways of displaying results collected along a transect. Some have already been described in previous sections whilst others are described here for the first time.

When to use

Most transects are lines or routes along which measurements/observations have been made. These measurements have been taken *either* continuously *or* at regular intervals. A transect is often used when a whole area cannot be surveyed and is therefore used as a sampling technique. In many cases several transects can be put together to make a type of incomplete pattern.

Examples

Changes in plant species (away from a footpath, up a hillside, etc.), land use changes with altitude, aspect, relief, etc., changes in building ages, styles and densities with distance from a town centre, changes in stone size and shape with distance down a scree slope.

Methods of construction

a. A summary of methods already mentioned

i. If the data is continuous or sequential use a simple **line graph**.

ii. If the data is not continuous then a **scattergraph** can often be the best method with distance along the transect as the horizontal axis. This can only be used if the other variable is quantifiable, ie. it is represented as a number and not a name (as in the case of plant names).

iii. If a visual impression is needed a **pictogram** type graph may be used to great effect. In this case it is the height of the figures that represents the observed data.

Worked example

Housing density with distance from the CBD

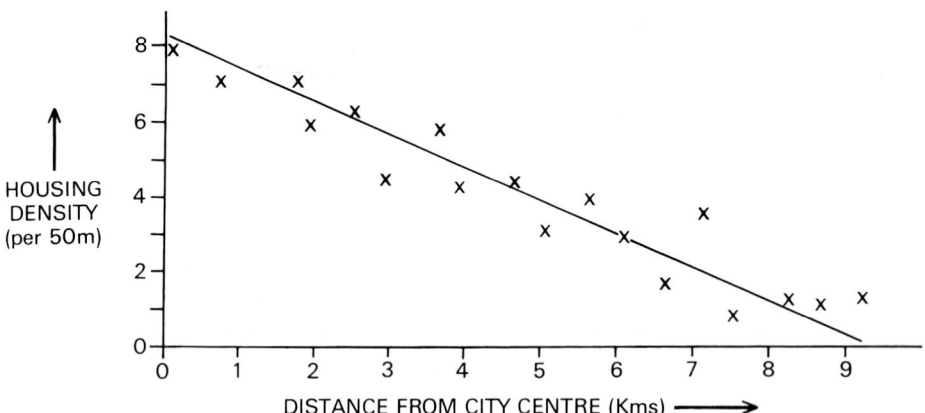

Worked example

Pedestrian densities (15 minutes survey) with distance from the city centre

iv. If **Profiles** or **Cross sections** are involved then a whole number of methods may be involved as described on pages 36 and 37. A profile or cross section is used as a base and other graphs are superimposed onto them. Types of graph include line graphs, barcharts, histograms, pictograms, pie graphs, rose diagrams, etc.

b. Other methods of showing transect data

1. MAPPING.

This is one of the commonest ways of displaying transect data and involves the initial mapping of the features under study such as land use, building ages, vegetation types, soil types, etc. Colouring or shading is used for each of the categories and they go directly onto the base map. Often it is possible to combine several transects to produce a more complete picture of the study area, for example several transects radiating out from a town centre showing such features as function, housing age and density, environmental quality, etc. The more transects the more accurate the overall impression your results give although it must be stressed that transects are a type of sampling and only by covering all the area will you get a complete picture of reality.

Worked example

Building ages along a street leading away from the centre of a small village

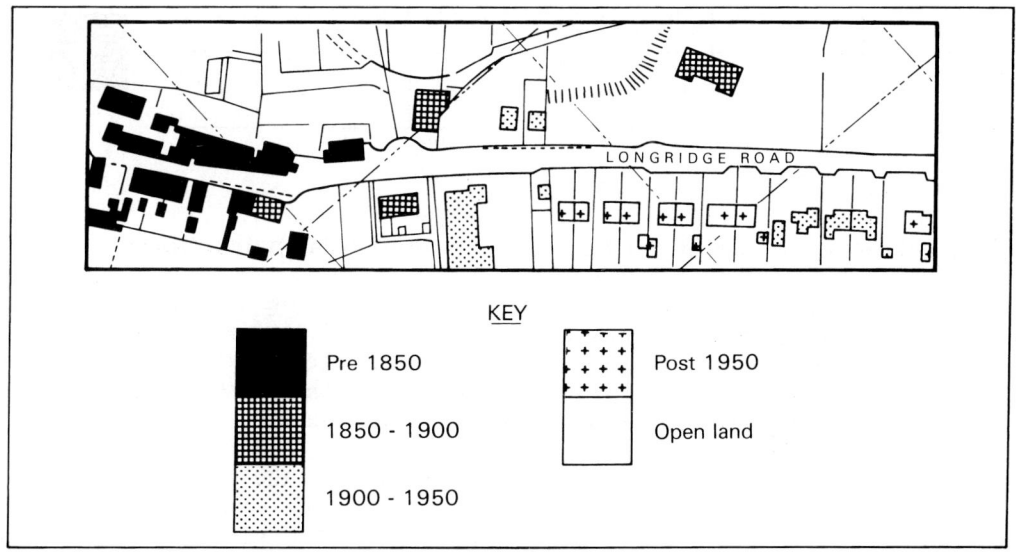

2. BARS.

These can be used in two ways.
a. To show information already displayed on a map. Here bars are used to simplify information on the map. Blocks or bars represent the length of the transect and are then divided according to the size of each of the categories present. Measurements are taken directly off the map. Areas that are adjacent to each other *but* have the same category are put together into one section (see the worked example below). In this way individual buildings are not shown.

This method 'straightens out' transect lines and often makes the identification of zones more obvious.

Worked example

The same information has been used as for the example above and therefore the same key applies.

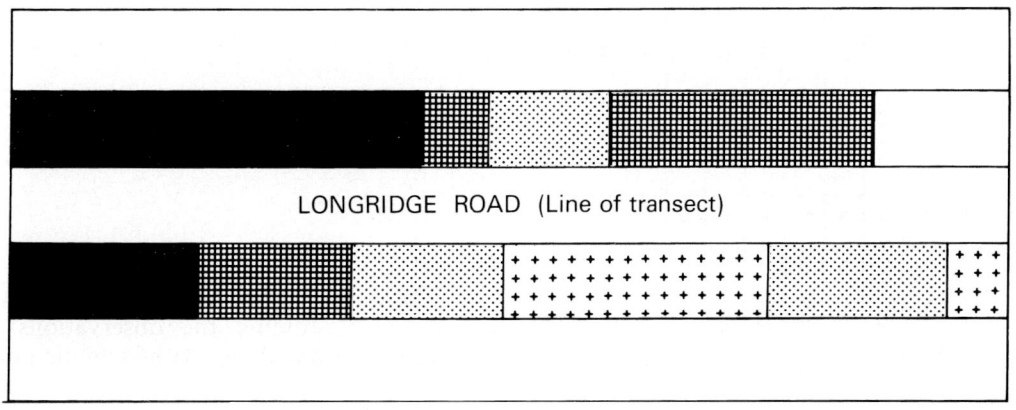

b. Bars can be drawn when a single measurement/observation has been made at regular intervals along a transect line.

A transect was taken up a hillside and at regular intervals (every 50m) the land use was noted. Other relevant data such as soil variables, slope angle, aspect, altitude, etc. can be included below the diagram.

Worked example

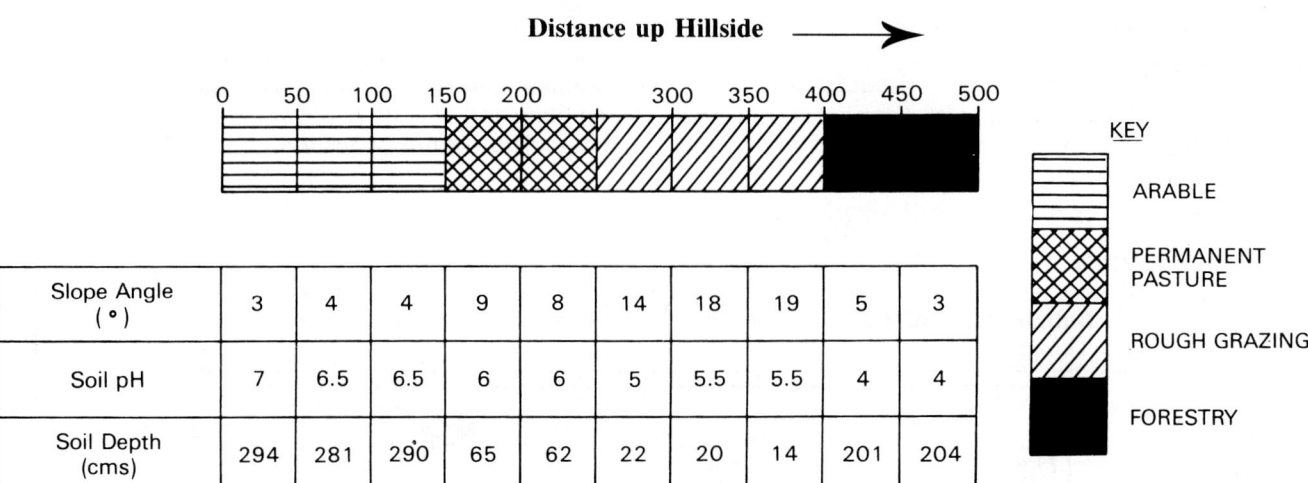

Distance up Hillside ⟶

Slope Angle (°)	3	4	4	9	8	14	18	19	5	3
Soil pH	7	6.5	6.5	6	6	5	5.5	5.5	4	4
Soil Depth (cms)	294	281	290	65	62	22	20	14	201	204

KEY
ARABLE
PERMANENT PASTURE
ROUGH GRAZING
FORESTRY

This type of diagram can be incorporated into profiles and cross sections. The main weakness of this method is that there is no guarantee that a single sample at each point will reflect reality (the sample size is too small). Several transects drawn parallel to each other may give a clearer picture and even show possible zonations. The next two methods are more reliable methods of graphical representation.

3. KITE DIAGRAMS.

These are very useful diagrams when it comes to showing *several* observations that have been made at each of the sample points along the transect. In this way they are a more sophisticated version of the method just described.

In the worked example below 10 plants were recorded at metre intervals along a transect that led away from the edge of an eroded mountain footpath. The plants were sampled using a transect line and a simple point frame and gave the following results:

Distance from edge of footpath (Mˢ)											
	0	1	2	3	4	5	6	7	8	9	10
BARE EARTH	8	2	0	0	0	0	0	0	0	0	0
MOSSES	2	1	0	1	2	0	2	1	2	0	2
MOOR GRASS	0	6	8	4	0	0	0	0	0	0	0
SEDGES	0	1	2	5	3	0	0	0	1	4	1
BILBERRY	0	0	0	0	4	8	5	1	1	2	6
HEATHER	0	0	0	0	1	2	3	8	6	4	1

Method of construction

1. A separate row is used for each of the individual species found. In this case there are six rows and each row must be ten units *wide* as ten plants were sampled at each site. The *length* of each row represents distance along the transect.

2. At *each* of the sample points the width of the constructed 'kite' represents the number of times that particular plant was observed, for example at the sample point that was 1 metre away from the footpath, moorgrass was observed six times and so at that point the moorgrass row is six units wide.

The widths should be centralised at each sample site around a central axis as shown in the following diagram:

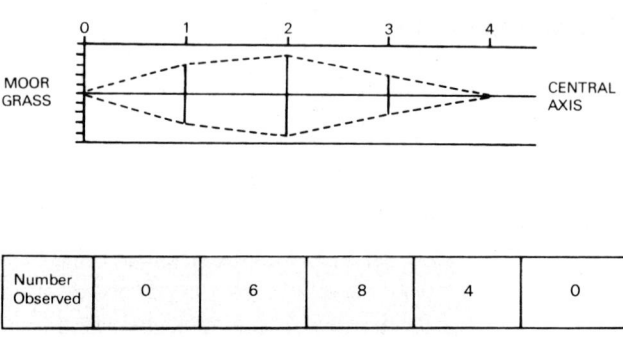

Number Observed	0	6	8	4	0

3. These lines are then joined to give the 'kite' appearance. Joining the observations in this way suggests gradual change from sample point to sample point.

4. Single shading or the use of different colours for each row can prove visually attractive. Once again information may be included either as a table or as another form of graph (line graph, barchart, etc).

Kite diagrams show not only where the various individuals occur but also the relative frequencies of each. They can give a good visual impression of any spatial change and any associations between the individual species and/or related variables.

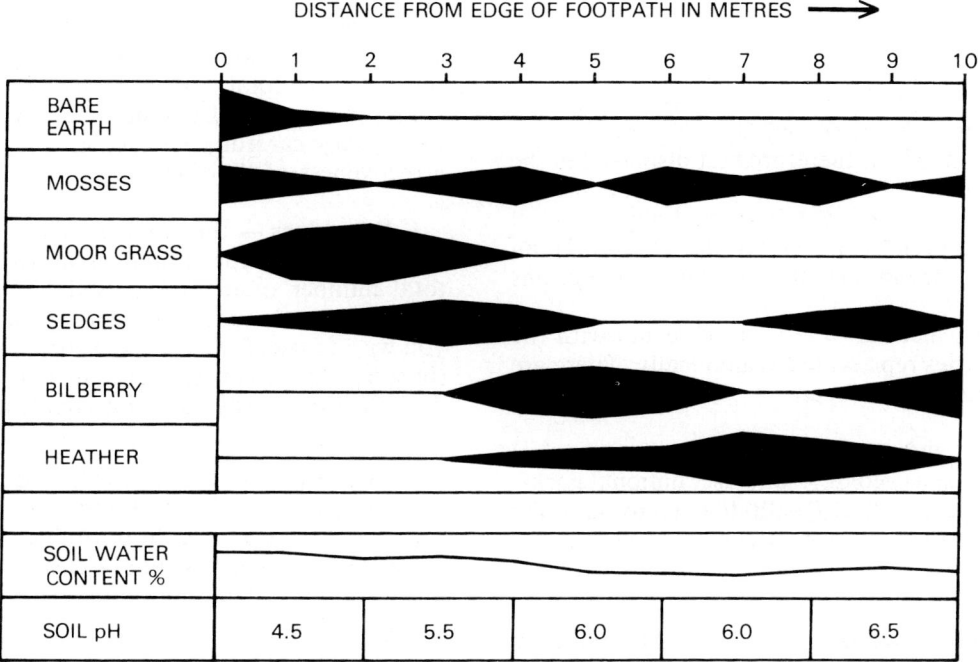

4. BLOCK GRAPHS.

These are very similar to kite diagrams as they also show the change in the observed variables with distance along a transect. There is one important difference, block graphs use observations that have been made *continuously* along the transect rather than at regular *intervals*. Because of this they must give a more accurate picture of reality.

Worked example

An urban transect starting near the peak land value and finishing near the periphery of the CBD was set up. The transect was then split up at regular intervals (every 50 metres) and the functions of the premises along the transect noted and put into categories. Each shop frontage was measured and noted.

Each category size is converted to a percentage of each 50 metre stretch. Each 50 metre stretch is represented therefore as 100% and then consequently subdivided into the components found within that 50 metre stretch. In this respect this type of graph could be said to be a histogram version of the composite barcharts described on page 21.

A CBD transect showing changes in land use

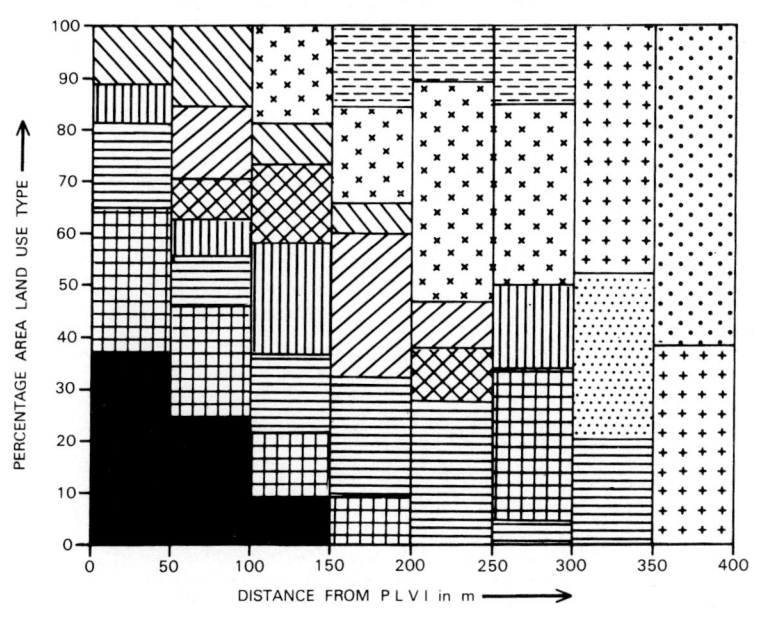

Representing information on maps

When to use

In many cases the information collected in the field will show some form of spatial distribution. In these cases the information is therefore best displayed in the form of a map rather than a graph. The use of different types of map is a basic technique for all fieldworkers especially as visual description is an important first step in seeking to explain distributions of this type.

A map is a scaled down version of reality with the various properties represented symbolically. There are *two* major categories of map.

i. Maps that show variations in non numerical variables such as geology, land use, national parks, etc. These are called **Qualitative maps** and are relatively straightforward. Because of this they are not really dealt with in this booklet.

ii. Maps that show variations in numerical values (observed data) over a given area. These are called **Quantitative maps**. There are several different ways of representing such data on outline/base maps.

DOT MAPS

One of the simplest ways in which certain data may be shown on a map is by drawing dots to illustrate the distribution of the items, for example crop yields, number of people, cattle, shops, etc.

If the number of data points is small and their exact location is known then the dots are drawn as 'point patterns' (at the exact location where the data was collected).

In many cases however the information is only available for areas and *not* specific points. In this case it is necessary to produce an approximate spread of points. This spread will represent the desired density by the spacings between the resulting points.

It is rare for the data to be represented by a single dot. In this case dots must be assigned values that are related to the total observed data. The choice of these values (along with the actual size and distribution) has to be very carefully decided as they can affect the effectiveness of the final map.

i. Dot values — the number of items shown by each of the dots will depend on the scale of the map and the total number of items involved. As a general guide divide the total number of items by the number of dots you wish to use, for example if there are 300,000 items (people, sheep, cattle, etc) and you wish to use 500 dots then obviously each dot equals the value of 600 items.

ii. Dot size — in this particular case the best effects are brought about by trial and error or practice and experience. If too large the dots overcrowd the map, if too small then they may not be clearly visible. Areas of greatest observed density should have dots that appear to be merging into each other.

iii. Dot location — if the information is to represent areas rather than specific locations then dots must be spread evenly throughout that area. Dots must be placed close to any boundaries (fields, parish boundaries, etc.) otherwise these areas may stand out as 'empty areas' on the map which is not the case in reality.

Colour can be used to show the mapping of several different categories on one map in order to compare/contrast their distributions, for example the mapping of different ethnic groups in an urban conurbation. Different symbols may be used just as effectively.

Dot maps can be drawn quite quickly and are easy to produce if thought out carefully beforehand. They give a good general impression of changes in density from place to place but they have a low level of accuracy once large dot values are used.

Worked examples

1. Schools using Hothersall Lodge Field Study Centre academic year 1986 – 87.

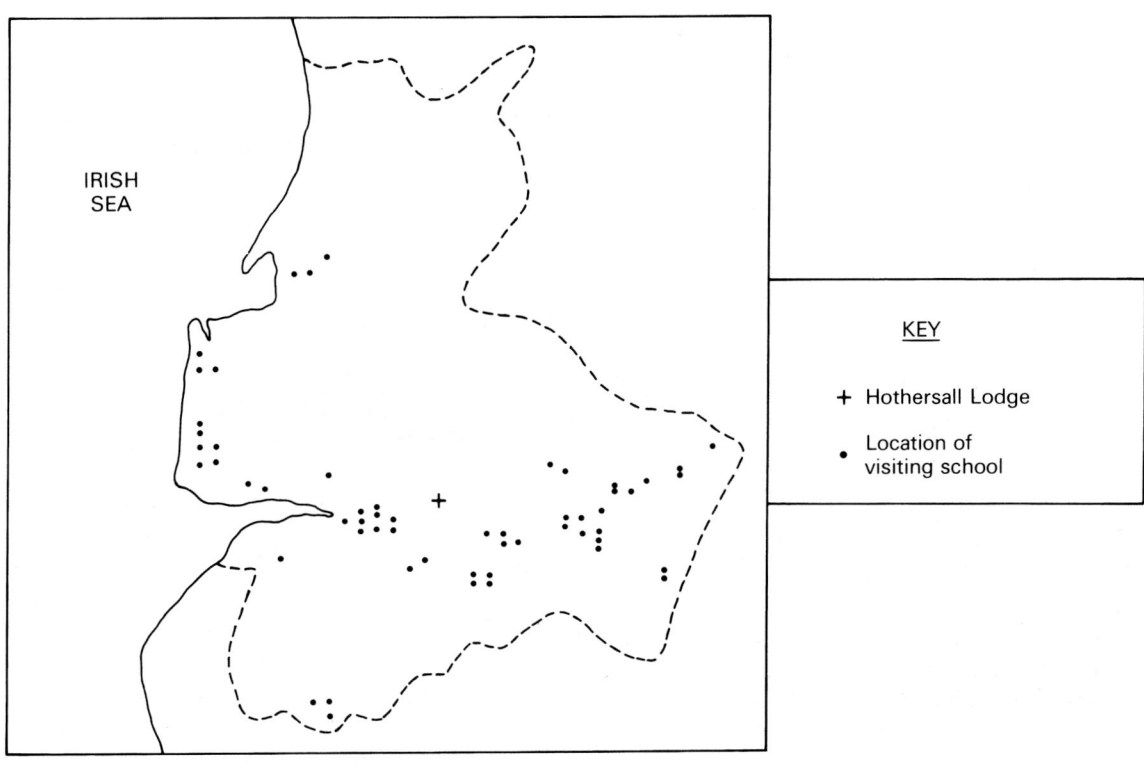

IRISH SEA

KEY

+ Hothersall Lodge

• Location of visiting school

2. Limestone pavement distribution (after Ward and Evans). (Source: *Teaching Geography* July 1982).

200 km

Peak District

Limits of last glaciation

Mendips

3. Distribution of blackface sheep in receipt of hill sheep subsidies in Scotland (Source: W.J. Carlyle *Geography*, January 1972).

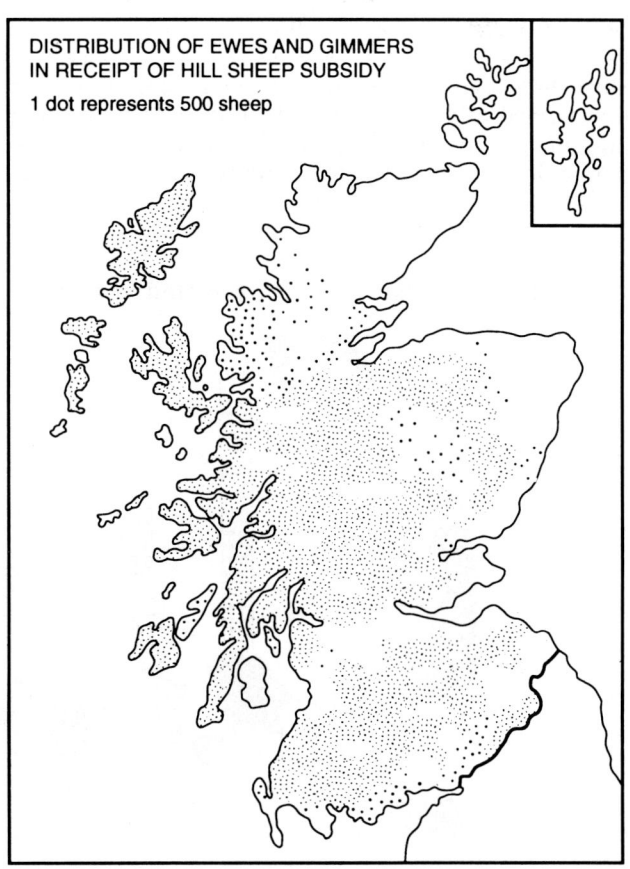

DISTRIBUTION OF EWES AND GIMMERS IN RECEIPT OF HILL SHEEP SUBSIDY

1 dot represents 500 sheep

Proportional symbols

In this particular case symbols are drawn proportional in size to the observed data. Appropriate types of symbol include bars, composite bars, circles, divided circles (pie graphs), spheres, squares, and pictograms.

The main difficulty with this technique is to choose an appropriate scale. One of the simplest methods is to use a scale line as shown on page 47. In practice the two most common methods use proportional bars and proportional circles. Because of this these two methods are described in detail at this stage.

i. Proportional bars — in this particular case the length of the bar is proportional to the observed value. If the bars are drawn too long they may overcrowd the map, if too small then the differences are not always easily discerned. The base of the bar is drawn at the location it is supposed to represent. Bars should be uniform in width.

Often the bars have component categories; if so they are subdivided in the same way as composite or multiple bars (see pages 20 and 21). If there are no component categories then the bars should all be shaded the same (usually black).

It is equally possible to use bargraph and histogram data on base maps to show spatial patterns. Remember shading must be uniform throughout and a key must be included.

Worked examples

During 1985 visitors to Fairsnape Fell access area were interviewed and their point of origin was noted and how many times they visited the area in a year. The map below was constructed to show the results.

2. Variation in wealth in 13 Hampshire villages, 1665. (Source: *Teaching Geography* July 1982).

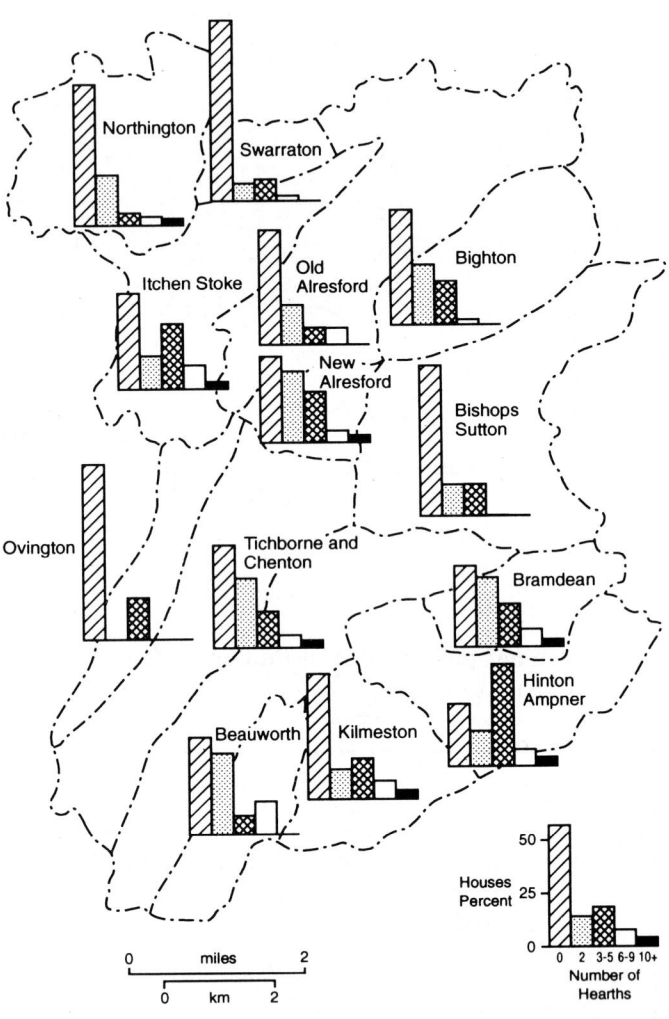

1. Visitors to Fairsnape Fell, Bowland in 1985

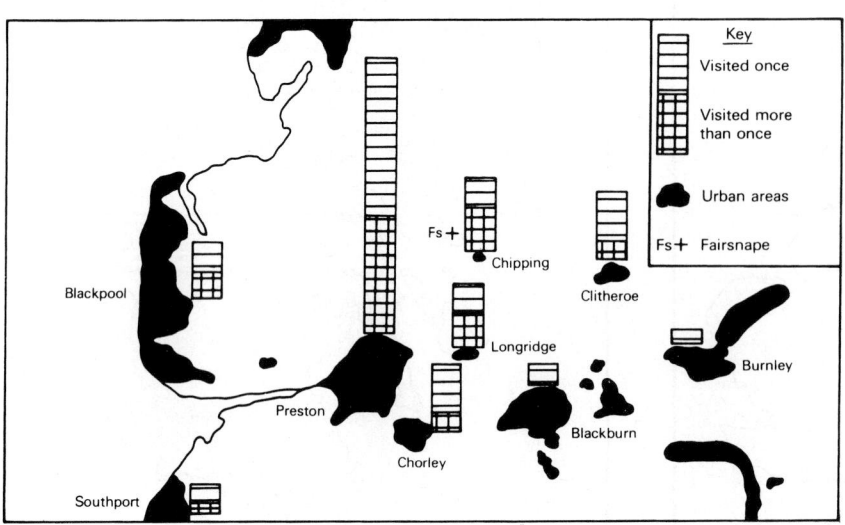

ii. Proportional circles and pie graphs — these are the most common types of symbol to be used on base maps and they can take several forms. The previous section on pie graphs discusses the problem of scale (page 22). To recap, the radius of the circles is not directly related to the observed value, it is the **square root** that has to be used once a suitable scale has been established. It is very unlikely that the square root of the actual observed value will provide the best circle size to fit onto the base map therefore a scale line is usually used. This method ensures that the optimum circle size is chosen for the scale of the base map. Circles that are too small fail to show enough detail whilst those drawn too large will either not fit onto the map or will badly overlap.

a. Constructing a square root scale
i. Draw a linear scale that will accommodate all the square root values of the observed data.

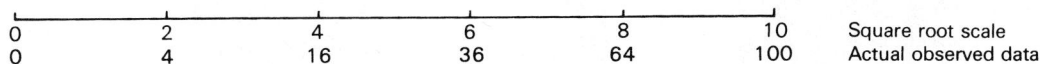

| 0 | 2 | 4 | 6 | 8 | 10 | Square root scale |
| 0 | 4 | 16 | 36 | 64 | 100 | Actual observed data |

ii. At the far right hand end of the scale (ie. the largest data point) draw the largest circle that will fit neatly onto your base map. From the centre of this circle draw a straight line connecting it to the left end of the linear scale line (see the diagram below).
iii. You have now constructed a scale line off which all the other circles' radii can be read. This is done by finding the square root value of the observed data along the base line and then by measuring the distance between the two lines as shown in the example below (an observed value of 52 gives a square root of 7.21, the radius of the circle is measured off as 6mms in this example).

All the resulting circles are proportional to each other whilst at the same time they are an accurate representation of the observed data.

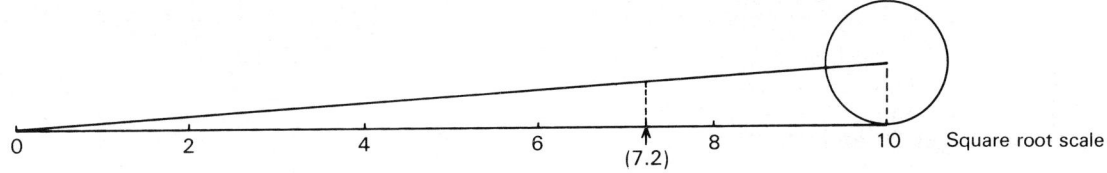

| 0 | 2 | 4 | 6 | (7.2) | 8 | 10 | Square root scale |

b. Drawing the circles on the base map
Start by drawing the largest value onto the map. Work away from this point fitting in the other circles around it. As far as is possible place each circle with its centre over the site where the data was collected. If it is necessary to overlap circles hide parts of one behind the other (see the worked examples). Even if only a small section is left showing it should be sufficient to convey the right impression. As such this technique is often useful in highlighting areas of high density.
NB. If pie graphs (divided circles) are to be used DO NOT OVERLAP.

If the data are shown as single/undivided circles omit all other details from inside the circles, for example boundary lines, rivers, roads, etc. In practice it is easiest to draw the circles then add the background details around them later.

Circles may be shaded in black or hatched but as a general rule, wherever possible leave them blank.

c. Displaying the scale
The first method is to draw a simple scale line as described above but the most common practice is to show the scale as a **nested set** of circles. The method of construction is basically the same as for scale lines. The optimum size is chosen for the largest circle and then the rest are constructed according to their square root value. Often a scale line is drawn first in preparation and then a sample number of circles are taken from it to make up the nested set.

Worked example

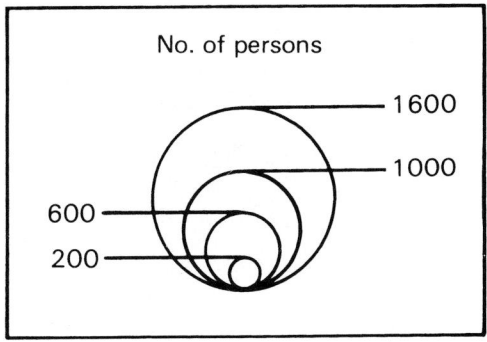

No. of persons
1600
1000
600
200

1. Regional distribution of unemployment in the UK, 1979 and 1983
Source: P.R. Mounfield, *Geography*, April 1984.

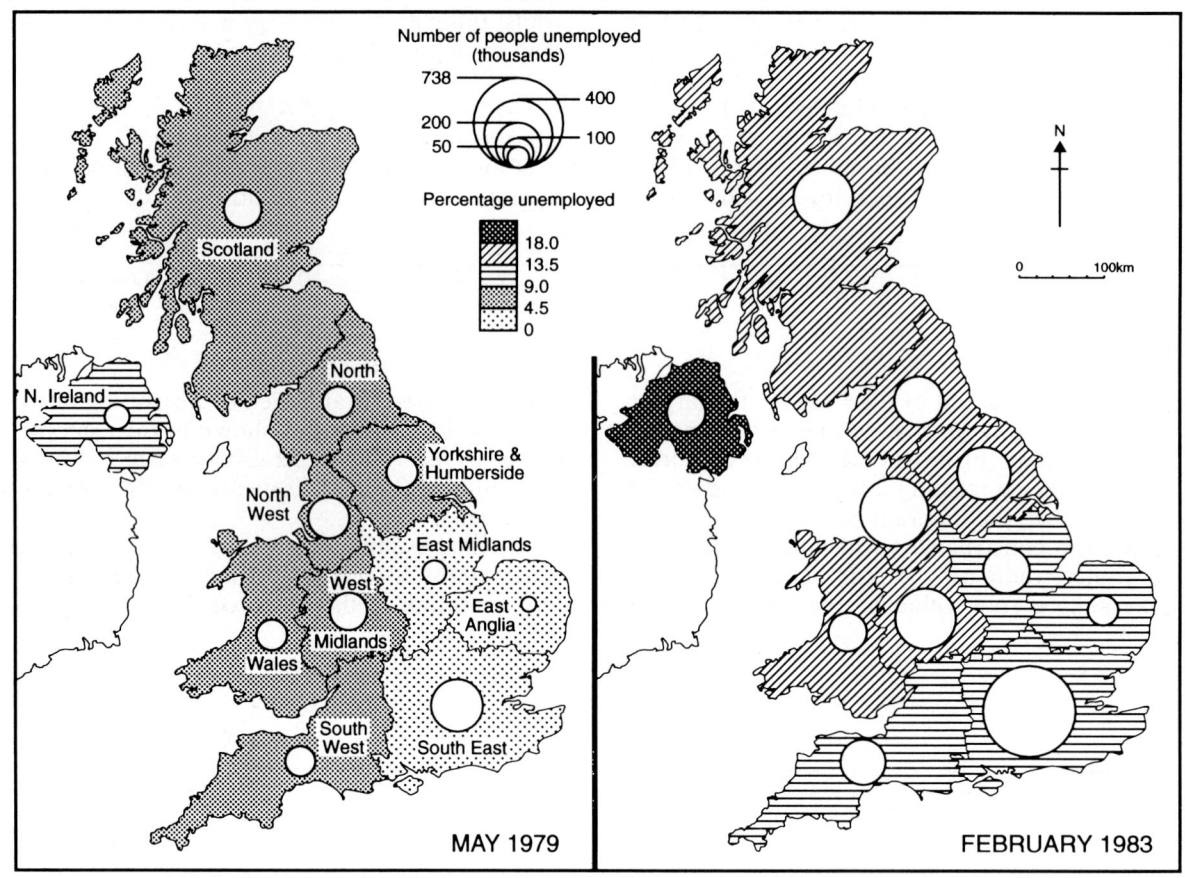

2. Cities in Latin America with populations of over 500,000
Source: P.R. Odell, *Geography*, July 1974.

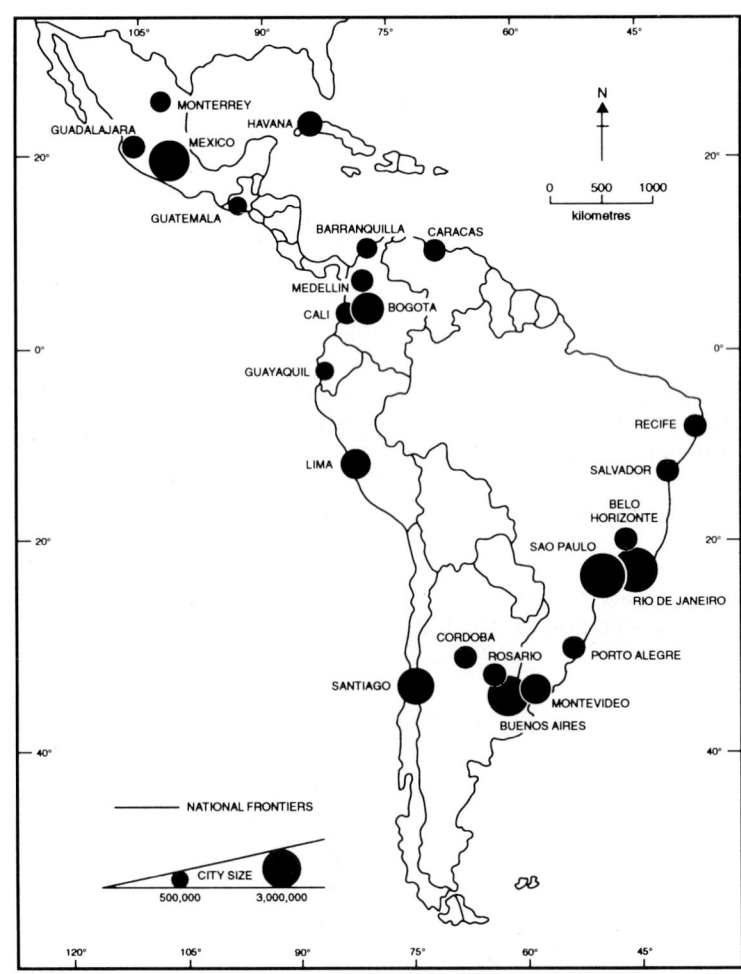

Pie Graphs/Divided Circles

Sometimes it is desirable to divide the proportional circles, ie. when the observed data has definite component categories. This additional information is easily displayed by dividing the circles into segments that represent the relative proportions of these component categories, for example a map showing population may divide the circles according to male/female categories or into age categories.

The method of dividing the circles is explained in detail in the pie graph section on pages 22 and 23.

Worked examples

1. Employment structure in Portugal
Source: Lewis and Williams – R.L. King, *Geography*, July 1982.

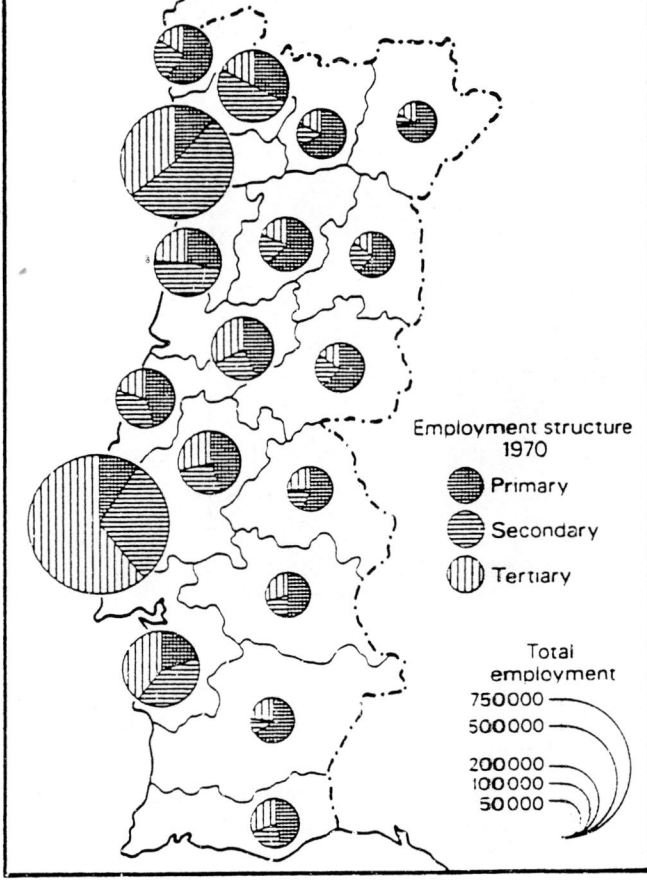

2. Markets for coal by region in 1977
Source: D.J. Spooner, *Geography*, January, 1981.

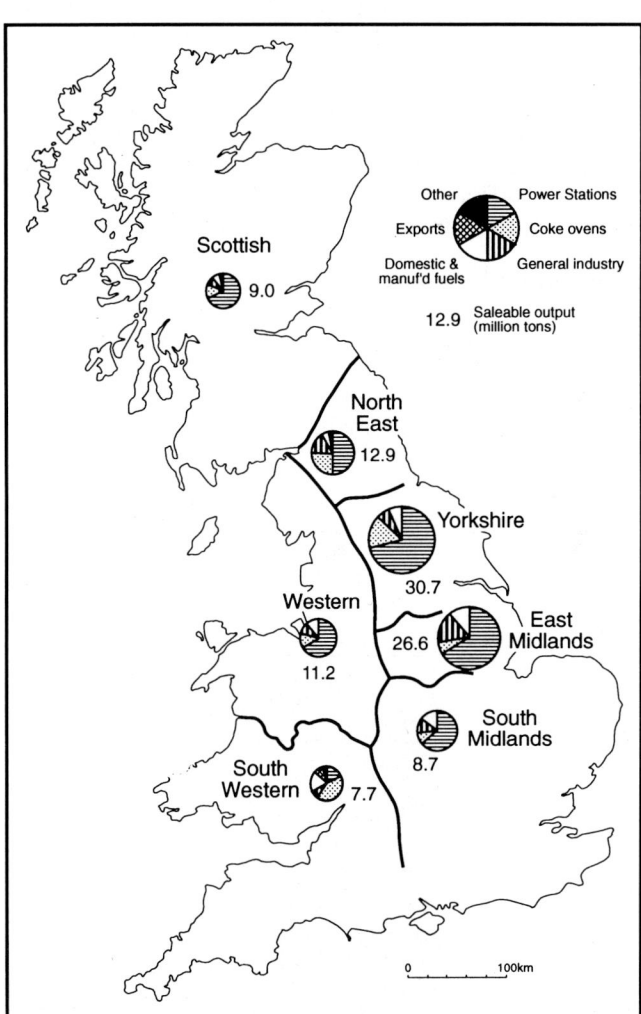

Proportional circles have an added advantage in that they use square root values. This means that a large range of values can be easily displayed. If drawn carefully they are visually attractive and are easily divided if necessary. Their main drawback is that they are time consuming to construct.

49

1. CHOROPLETH MAPS

Choropleth (shaded) maps are predetermined shading according to a key. The intensity of the shading gives an indication of the size of the data values.

There are two important considerations when using these types of map.

i. The information must be presented in a certain form. Different sized areas will be used on the map (counties, districts, fields, etc) and data must be in the form of proportions or expressed in terms of area or density. This eliminates the problem of the different areas being different sizes.

ii. The data has to be classified into a key. This key must use a logical and systematical method as the categories must reflect the nature of the data used. The different ways of classifying data into categories are discussed on page 11.

Method of construction

i. Select and construct a base map that shows the internal boundaries between the areas studied (fields, parishes, etc). The smaller these units the more accurate the map.

ii. Work out an appropriate key taking care that no value can appear in more than one category, for example categories should be 2 − 3.9, 4 − 5.9, etc. rather than 2 − 4, 4 − 6, etc. For the best visual results there should be more than 3 shading types but no more than 9 or 10. The shading should get darker the greater the value.

If colour is to be used it is best to use groups of colours from the spectrum such as:

Red	High Value	Violet
Orange	↑	Blue
Yellow	↓	Green
White	Low Value	Yellow

White may be included but it has several problems:

a. white is often used to represent areas where the data is unavailable;

b. it is impossible to tell whether a white area has been left blank in error or on purpose.

iii. Shade in the units according to the key. It is common practice where adjacent areas have the same type of shading to leave out the boundary between the areas. This is shown in the two maps below.

iv. Draw and label the key. It should include the shading, the range of values and if possible the number of each points in each of the categories.

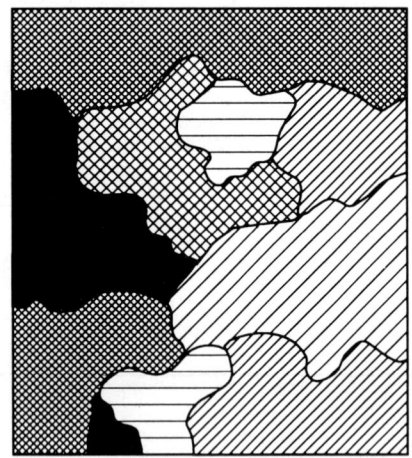

Limitations of their use

This type of mapping is easy to construct, these maps give a good visual impression of change over space and as such they are commonly used by geographers. They do, however, have several drawbacks.

a. They give a false impression of abrupt change at the boundaries of each area. This is unavoidable with this technique.

b. Any variations within each of the areas are not shown. Because of this smaller areal units are better than large ones.

Worked examples

1. Population changes, Lancashire District Councils, 1971 – 1981

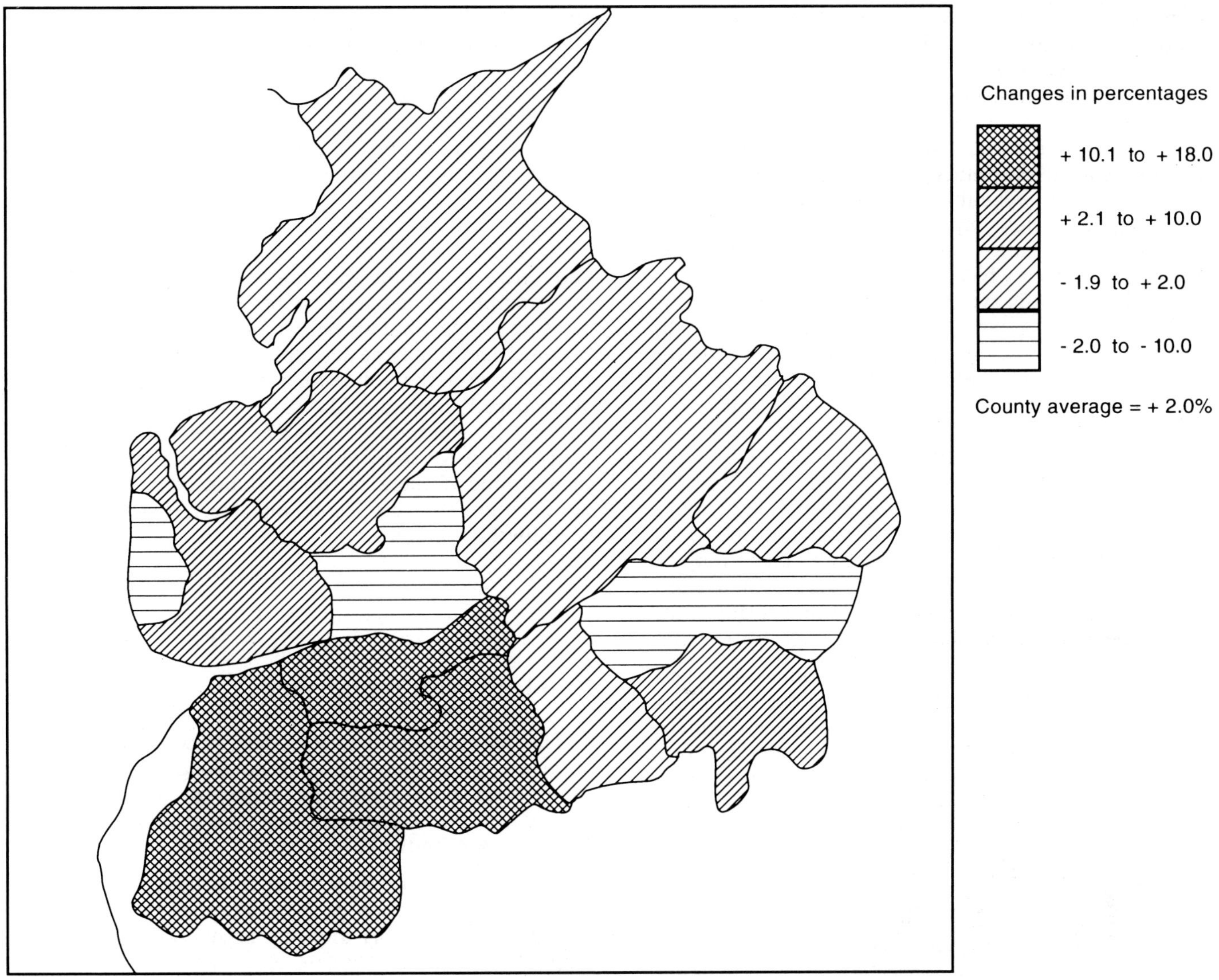

Changes in percentages
+ 10.1 to + 18.0
+ 2.1 to + 10.0
- 1.9 to + 2.0
- 2.0 to - 10.0
County average = + 2.0%

FRANCE
N
PORTUGAL
0
200
kilometres
Over 300
187.5
150
112.5
75
37.5
0

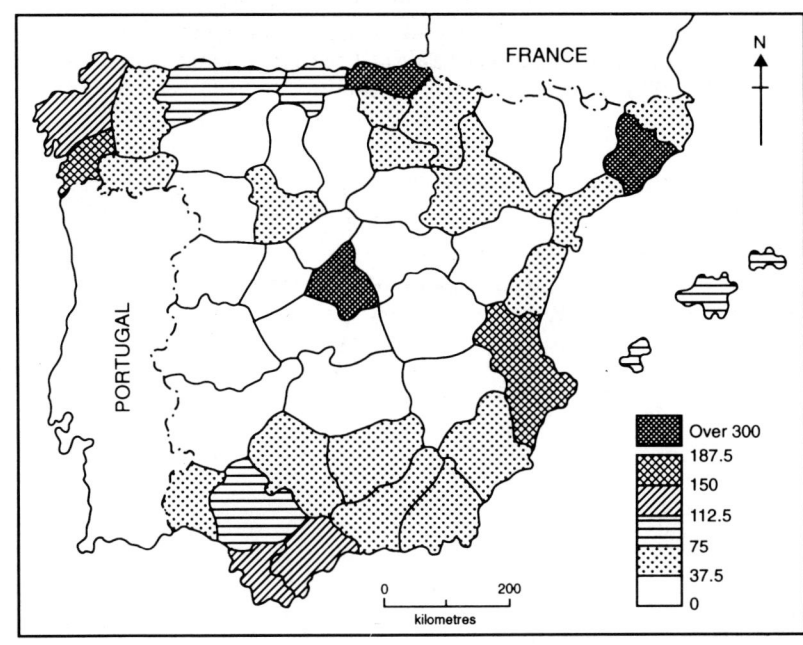

2. Spain, population densities per km², 1970
Source: H. Pullinger — *Teaching Geography*, April 1981

2. ISOLINE (ISOPLETH) MAPS

The method adopted here is to draw lines that join points of equal value. The best example is the contour line that joins points of equal height, but many other examples may be used: Isobars (air pressure), Isotherms (temperature), Isohyets (rainfall), Isovels (velocity). ISO means "equal", hence Iso (equal), therm (temperature), translates as "equal temperature".

This type of map is obviously at its best when data is measured at precise points or locations. Points may be chosen either at regular intervals or at random and plotted onto a base map. Once the lines have been drawn gradual change over space can be identified and the unreal effect which boundary lines produce on choropleth maps is avoided. For this type of mapping to be effective a large amount of data has to be collected and these maps are unsuitable where the data's distribution is patchy. There are other problems and these are discussed later on in this section.

Method of construction

i. Mark on the base map the observed values; make sure that they are located accurately at the exact position where the data was collected. There may be problems if areas are represented rather than exact points and this should be avoided. The more points the more accurate the end result.

ii. Decide on suitable values for the isolines. It is best to look at the range of values and then decide on how many lines will fit neatly onto the map between the maximum and minimum values. Usually a fixed interval is used, for example, every 10 metres on a contour map, but sometimes it may be more appropriate to look for natural breaks in the dispersion of points (see the section on classifying data on page 11).

iii. Construct the lines. This is the hardest stage as few lines will go directly through any of the observed data points. Because of this it is necessary to determine the values of intermediate points through which the isoline will pass. This is done by 'logical interpolation'. For example, if an isoline with a value of 20 is to be drawn between two points of value 19.5 and 21 then we assume it will be drawn ⅓rd of the way along in a line running **from** the lower value (19.5) **to** the higher value (21). This is shown clearly below.

Sometimes two or more alternatives may appear to be equally possible in which case it may be possible to estimate a 'missing' value by taking all the values around the missing point and working out the mean (average). This mean value is then located as a 'dummy' spot in a central position within the actual observed points and this may give an indication of which way the isoline goes. The other alternative would be to resample including more intervening points; this is only possible with certain data as values may change over time (pedestrians, stream velocities, etc.).

The construction of isolines is a skill that comes only with practice and trial and error. However, a prior knowledge of certain phenomena may prove helpful, for example isobars tend to be circular around low pressure systems but straighter between fronts.

Sometimes it may be desirable to shade or colour the spaces between the isolines in order to highlight change over space. The higher the value the darker or more intense the shading. This is the system adopted by many atlases for showing relief. Mark on the values of each isoline if there is room, if not, then put in a sample number (every second line, etc.). If shading has been used include a key.

Problems with isoline maps

i. The assumption is made that between any two observed values there will only be found intermediate values and never anything larger or smaller, ie. between the observed points there is a gradual change. This may not be the case in reality.

ii. These maps are only valid if a large sample is involved and collection of such data may be difficult. If only a small number of points are used the resulting map gives a false impression of accuracy.

iii. There is an element of personal judgement involved in the drawing of the lines. Two independent students using the same data may not produce identical maps using this method. There is a tendency to try to 'force' the lines where the data will not allow, just to complete a pattern that may appear to be emerging. This must be avoided.

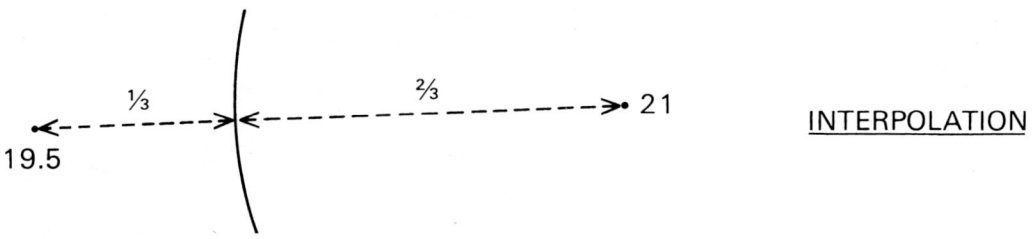

⅓   ⅔   21

19.5

INTERPOLATION

Worked examples

Example 1. Isovels on a river meander (River Langden)

Velocity readings were taken at regular intervals and depths across the meander. These were then plotted onto an accurate cross section of the feature and isovels were interpolated.

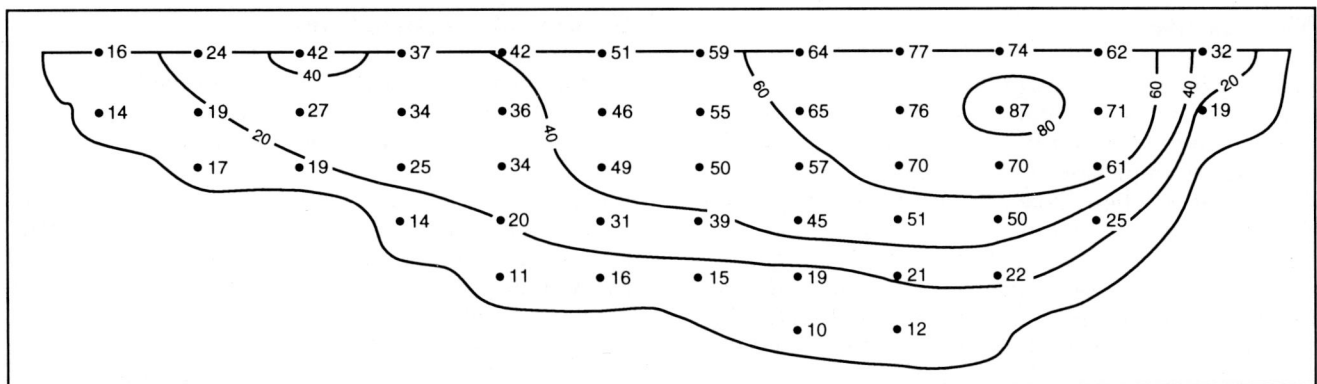

FIGURES IN METRES PER SECOND

Example 2. Pedestrian counts in Blackburn's CBD

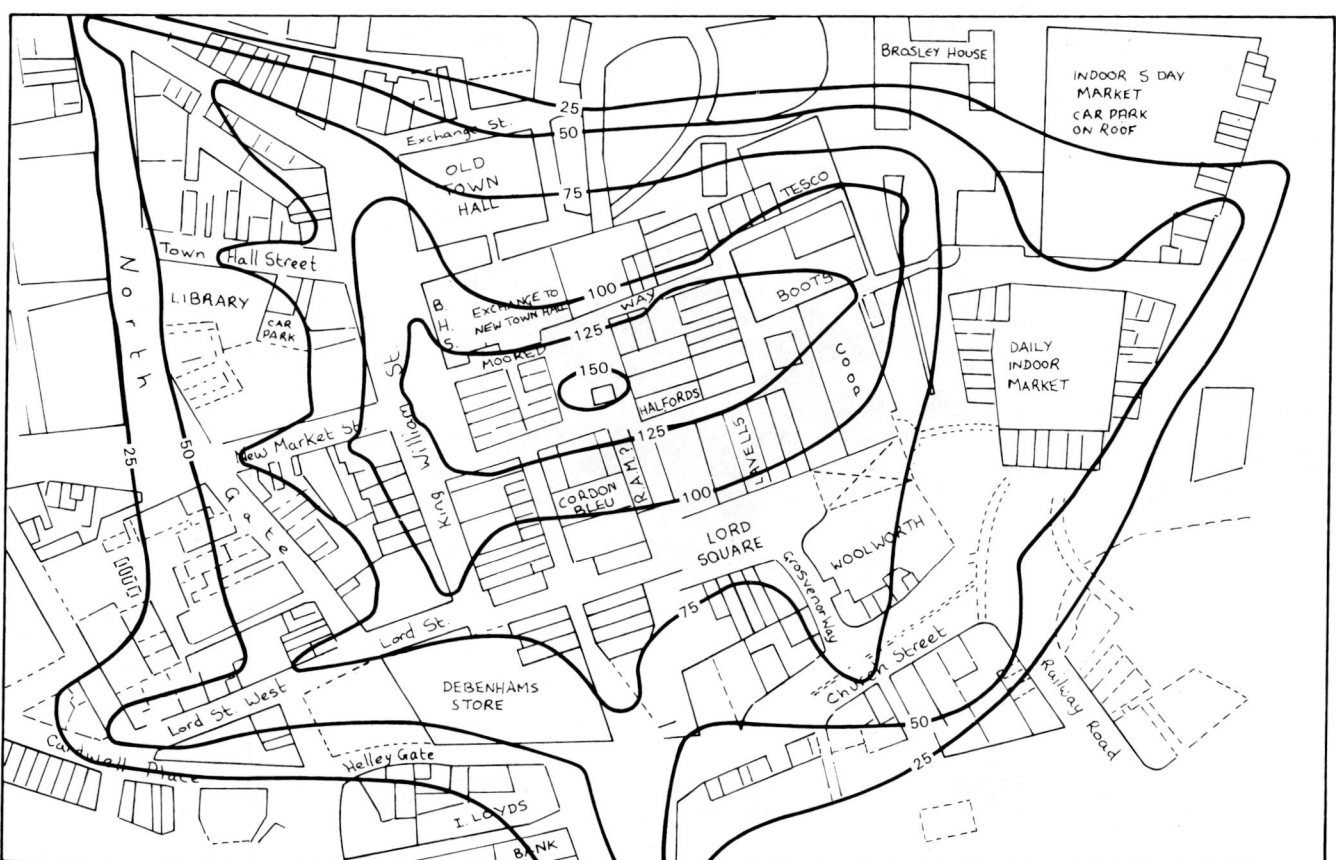

The Location Quotient

When to use

There are numerous ways of showing spatial variation of data, the most common being the use of different types of map. In some circumstances it may be desirable to emphasise certain characteristics of the observed data. This can be achieved by a combination of statistics and cartography (map construction).

The Location Quotient (LQ) is used to show degrees of concentration of one particular activity/characteristic within a chosen area, for example, how concentrated the coal mining industry of England is in the county of South Yorkshire.

Examples

Plant or animal species within a particular community: human activities (occupation, religion, ethnic group) within an area/settlement; shopping types within an urban CBD.

Method of construction

i. First work out the statistics. The final figure to be used (the locational quotient) is the ratio between the percentage of the group under study (within a given area) and the percentage of the whole group or population for that area.

If we take the example of rented accommodation, an area of town may have 28% of the town's rented accommodation but only 7% of the town's total housing, therefore:

the Location Quotient for that area =

$$\frac{\%\ \text{rented accommodation in the area}}{\%\ \text{total housing in the area}}$$

$$= \frac{28}{7} = \frac{4.0}{}$$

A location quotient is worked out for all the other areas under study.

ii. The results are put onto a choropleth map by the usual method (see pages 50 and 51). Don't forget the key.

Worked example

The location quotient for rented accommodation in the Borough of Greater London — 1974

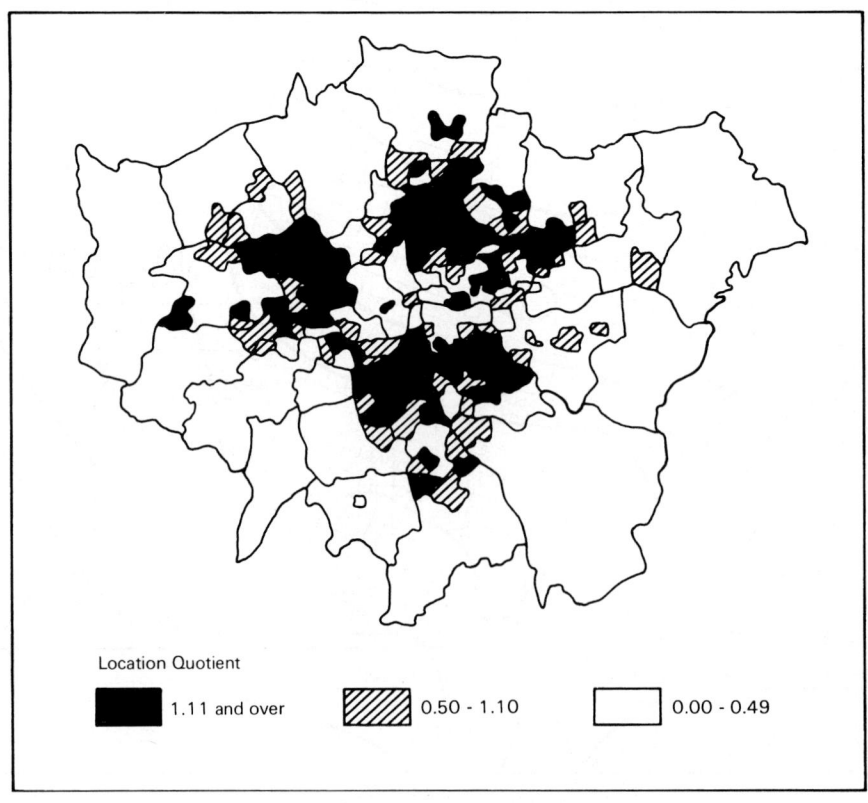

Location Quotient

| ■ 1.11 and over | ▨ 0.50 - 1.10 | □ 0.00 - 0.49 |

In theory the location quotient ranges from 0 to infinity. If an area has a value of more than 1.0 then the subgroup (rented accommodation) is over-represented within that area. The larger the value the greater the concentration of the sub group. In the worked example the areas shaded in black are over-represented, the areas in white are under-represented and those in line shading are on the borderline.

Mapping movements

When to use

Most of the methods already described refer to information that has been collected at definite, stationary points. Sometimes the information we wish to show is not static but mobile. There are several methods that can be used, according to type and suitability of the data collected.

Examples

Traffic and pedestrian flows, human and animal migrations, diffusion of innovations and ideas, movement of goods and services.

Methods of construction

As already mentioned there are several different methods the suitability of each depending on the nature of the data.

1. COMPOSITE BARS

The principle is the same as for the composite barcharts described on page 21. The length of the bar represents the **total volume** observed and is subdivided into its component categories according to their relative proportions. The arrow head indicates the direction of the movement involved (in the worked example below traffic).

The selection of an appropriate scale is important and at times it may prove a little difficult. The separate components may be shown as colours or shadings and a key may be used as it is unlikely that labels can be fitted onto the base map.

Worked example

Traffic flow at Stonebridge roundabout, Longridge

A detailed traffic survey was conducted to try and identify traffic flow around a well known local 'trouble' spot. It was carried out between 9.00 and 9.30 am. on Thursday, 10th January, 1988.

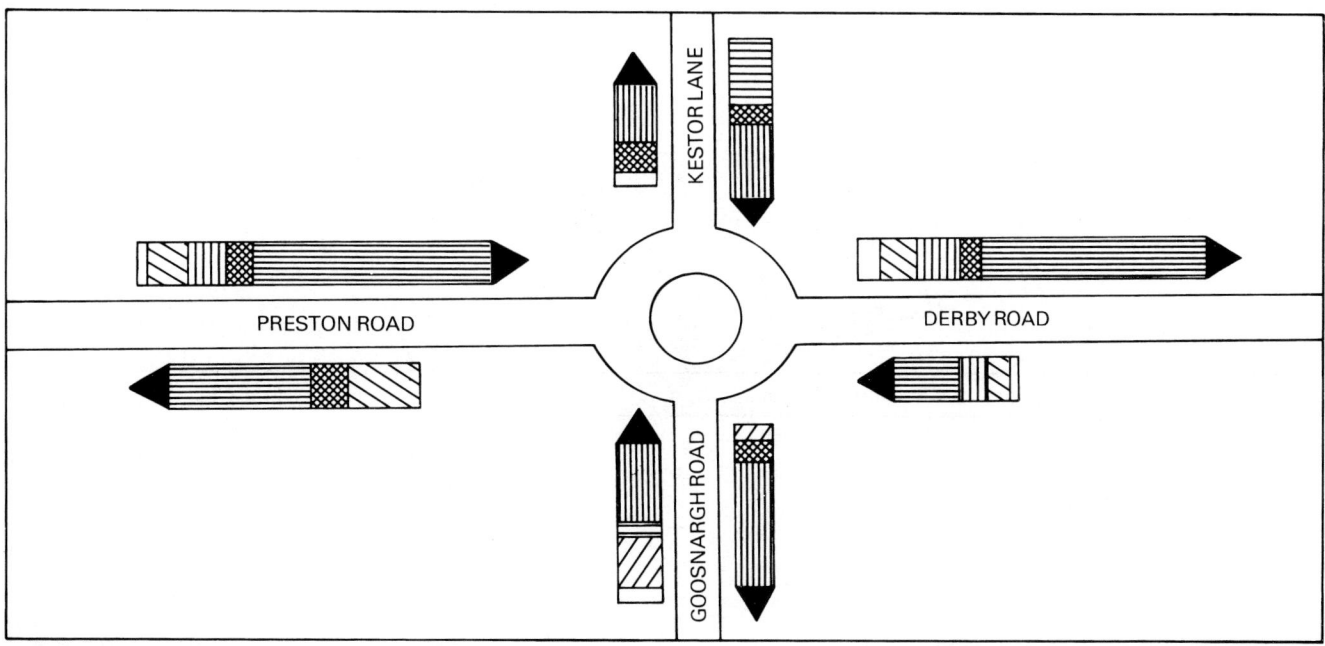

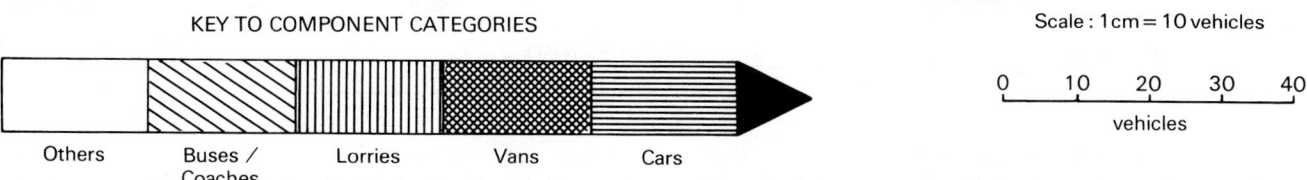

KEY TO COMPONENT CATEGORIES

Others   Buses /   Lorries   Vans   Cars
          Coaches

Scale : 1cm = 10 vehicles

0   10   20   30   40

vehicles

2. FLOW LINES

These are used when we are only interested in the **total volume** of the flow and not any of the component category values. In this case it is the **width** of the line that represents the total volume observed and not the length as in the previous example. Flowlines are at their most effective when there is only one type of information involved, for example, pedestrian flows, etc.

The most difficult part of this technique is the working out of an appropriate scale that will cover all the observed values. There are three alternatives.

i. A simple proportional scale where the width of the line is directly related to the observed values.

ii. A more complex proportional scale where the square root or the logarithmic value of the observed data is used. This is often necessary if the range of values is high which makes the above type of scale impractical (see the section on logarithms on pages 34 and 35).

iii. A graduated scale. Here a number of fixed line widths are used and the data values are categorised according to these widths.

The diagrams below show these three types of scale; as can be seen the first method is the only one that can be used easily for quantitative interpretation (reading values off the finished map).

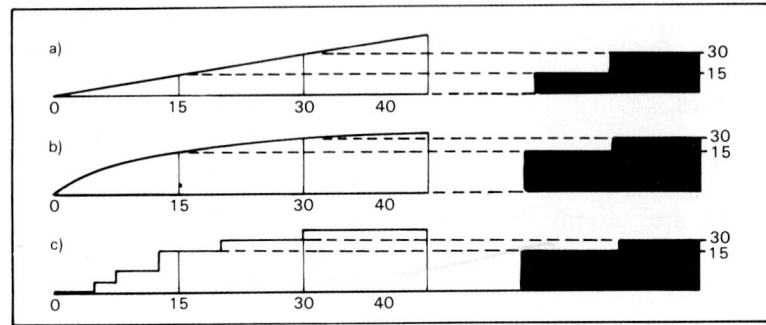

Note. For the first two methods changes are shown as a gradual change in the width of the line *but* if the third method is used (predetermined category sizes) then the widths of the lines have to change abruptly halfway between the sample points. This gives the visual impression of abrupt changes in flow rather than gradual changes which may often be the case. The examples below show these differences.

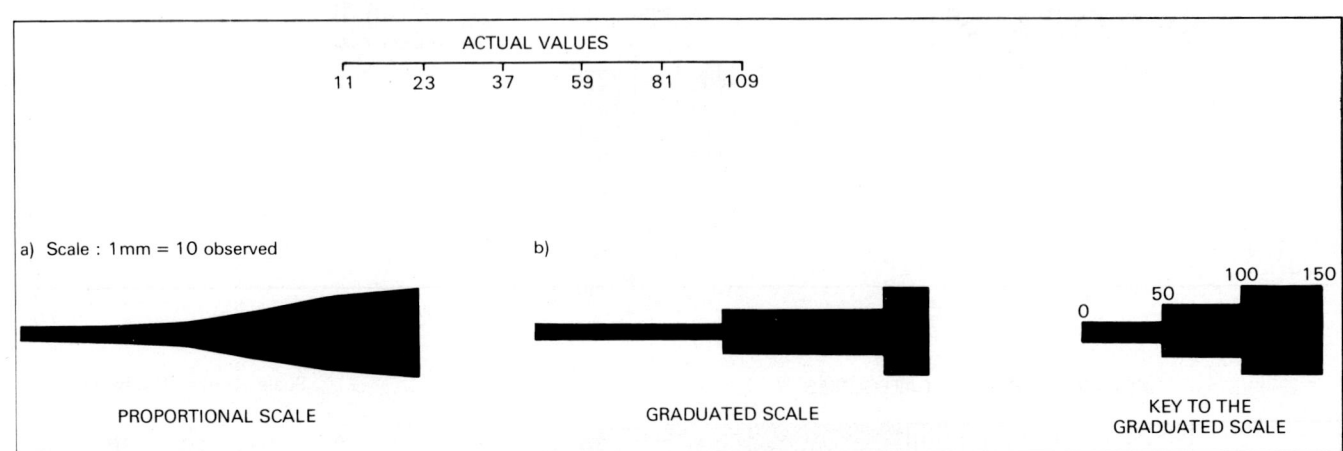

ACTUAL VALUES

11   23   37   59   81   109

a) Scale : 1mm = 10 observed   b)

PROPORTIONAL SCALE   GRADUATED SCALE   KEY TO THE GRADUATED SCALE

Shading is almost always in black to highlight the flow. If a graduated scale is to be used then class differences do not have to be uniform but the width of the actual lines must increase steadily by equal amounts. It is a good idea to keep the number of size classes small.

Worked examples

Traffic flow using a graduated scale.

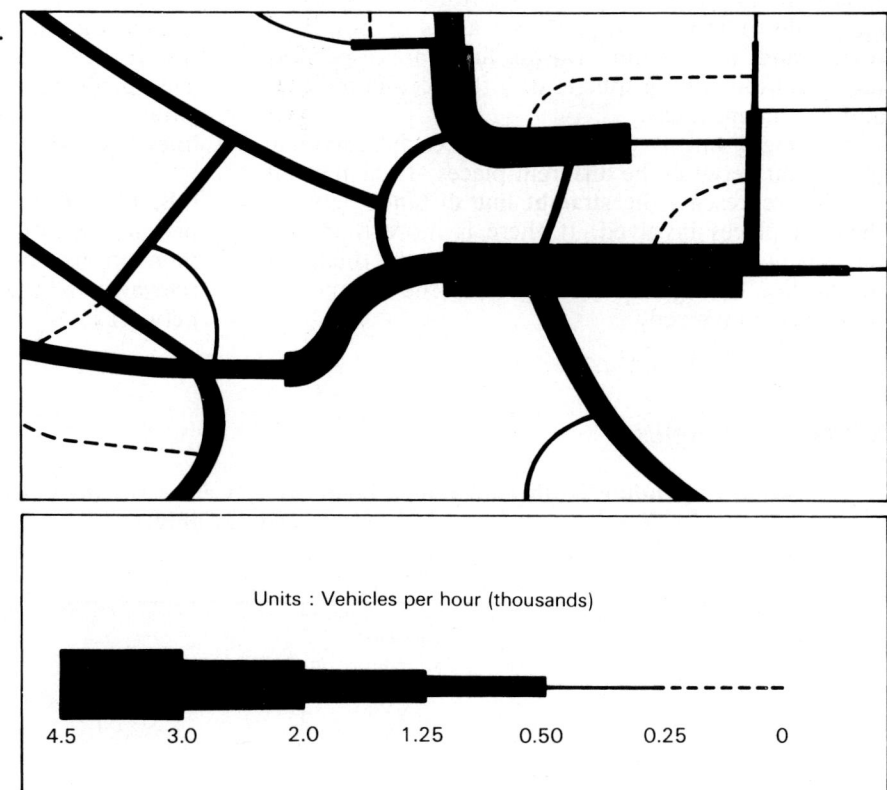

Units : Vehicles per hour (thousands)

4.5   3.0   2.0   1.25   0.50   0.25   0

Two directional flow. Sometimes it is desirable to show two-directional flows, eg. traffic flowing in different directions along the same route. Flow line maps can be easily adapted to show this data. The following example shows the pattern for pedestrian flows in a modern shopping centre. A simple proportional scale has been used since the range of values is relatively small. Two flow lines are drawn for each routeway observed and simple arrow heads are used to indicate the direction of flow.

The flow lines themselves have to be broken up into separate sections in order to eliminate the overlapping that would occur at junctions. This will simplify the overall appearance of the finished map.

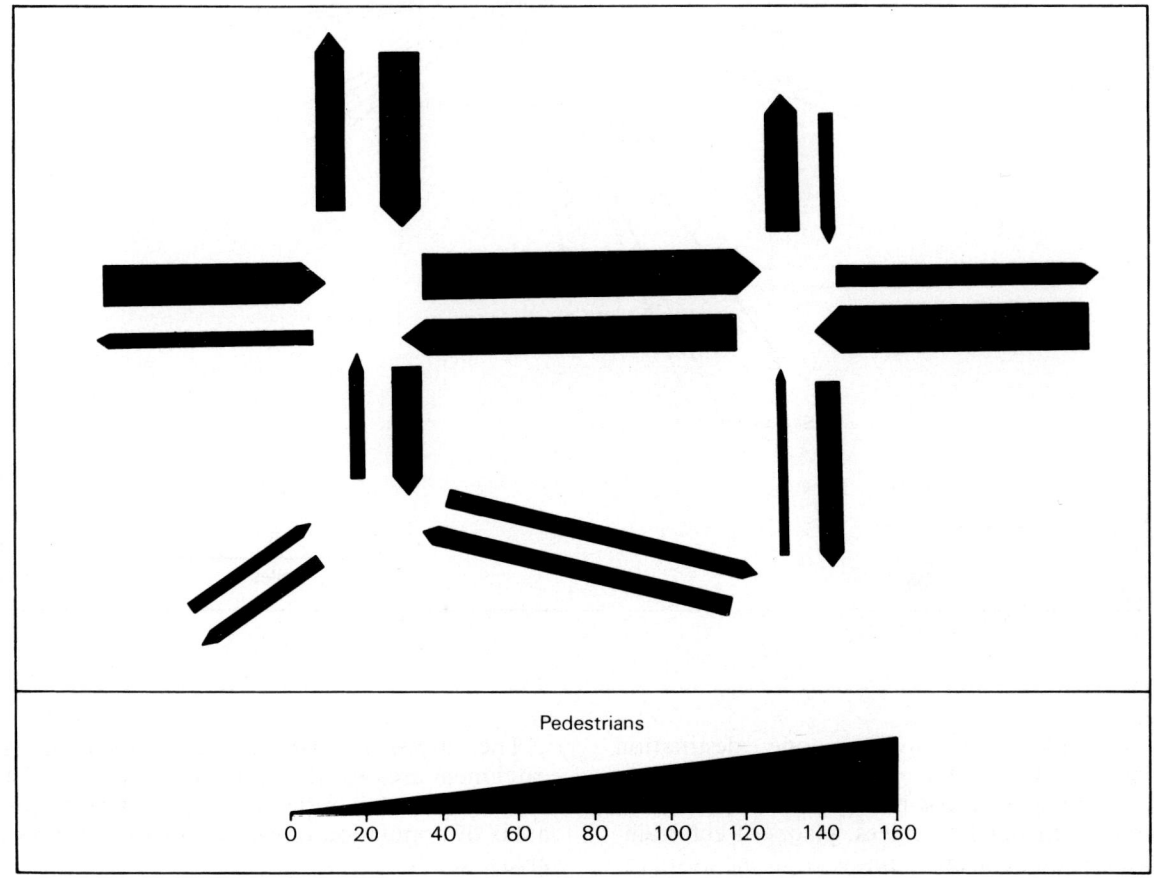

Pedestrians

0   20   40   60   80   100   120   140   160

3. DESIRE LINES

This is a simple technique to illustrate the direction and scale of movements between areas or points of origin and destination. Desire lines are especially useful for illustrating spheres of influence, hinterlands and catchment areas.

A straight line is drawn to show each individual movement between the different places. The length of the line represents the **straight line distance** between the two places involved. If there is more than one movement between the two points then the thickness of the line can be used to represent the number of movements involved.

These maps can become very crowded and complex if large amounts of data are involved. This is especially true if there is just a single destination point involved. This problem may be overcome by drawing a close circle around the destination point and excluding all lines that fall within that circle.

NB. Flow lines do not just involve the movement of people, goods, etc. but also the movement of innovations, ideas and information, and as such are relevant to the understanding of communication networks.

Worked examples

Schools attending Hothersall Lodge Field Study Centre for the academic year 1983 − 84 (residential courses only)

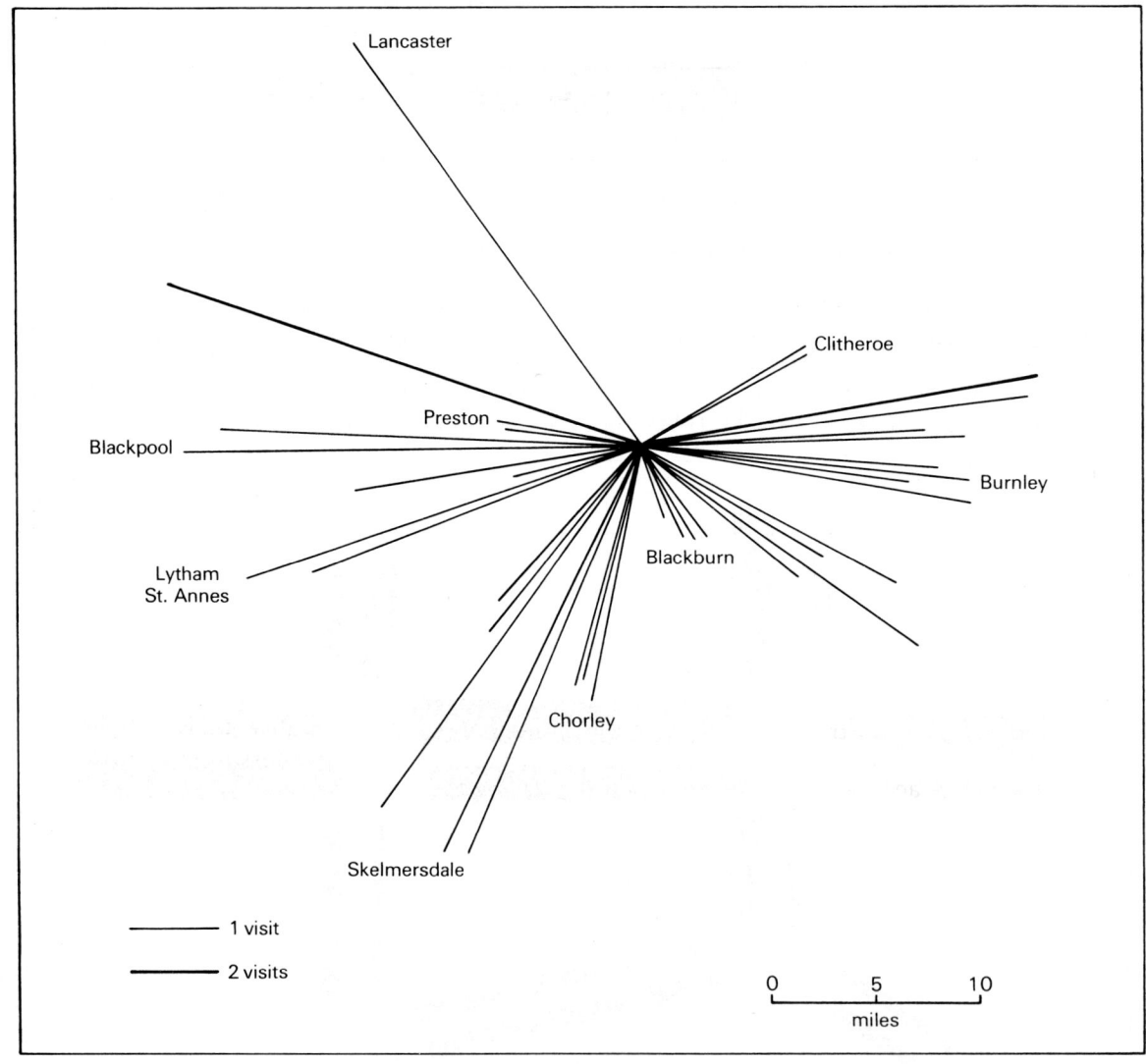

This example only involves one destination (Hothersall Lodge). More complex diagrams are possible where several destinations are involved, for example different service centres, places of entertainment, schools and colleges, etc.

The above diagram reflects Hothersall Lodge's catchment area for the academic year 1983 − 84 and to a lesser extent it reflects the influence of the larger areas of population on the number of schools in such areas.

Appendix 1

Table for working out the percentages of component categories in preparation for constructing composite barcharts.

$$\text{Equation} = \frac{\text{Percentage of each}}{\text{component category}} = \frac{\text{Actual number in category}}{\text{Actual total number}} \times 100$$

Cars sold in Clitheroe (1972 – 76)

| YEAR | COMPONENT CATEGORIES OF TOTAL DATA | | | | | | ACTUAL TOTAL |
| | FORD | | LEYLAND | | VAUXHALL | | |
	Actual	Percentage	Actual	Percentage	Actual	Percentage	
1972	22	$\frac{22}{67} \times 100 = 32$	30	$\frac{30}{67} \times 100 = 46$	15	$\frac{15}{67} \times 100 = 22$	67
1973	33	$\frac{33}{71} \times 100 = 45$	24	$\frac{24}{71} \times 100 = 30$	14	$\frac{14}{71} \times 100 = 25$	71
1974	44	$\frac{44}{79} \times 100 = 56$	20	$\frac{20}{79} \times 100 = 25$	15	$\frac{15}{79} \times 100 = 19$	79
1975	49	$\frac{49}{84} \times 100 = 58$	18	$\frac{18}{84} \times 100 = 22$	17	$\frac{17}{84} \times 100 = 20$	84
1976	50	$\frac{50}{86} \times 100 = 57$	16	$\frac{16}{86} \times 100 = 16$	20	$\frac{20}{86} \times 100 = 23$	86

The percentage worked out can now be drawn on a percentage composite barchart (see page 21). The bars will be drawn all the same length and therefore do not reflect the overall total trend, ie. overall car sales increased over the time period. What the graph does show is an accurate picture of which make of car increased its sales and which suffered a decrease.

Throughout this booklet these figures have been used three times. Which do you think is the most appropriate method? Does it depend on what you are trying to show?

Appendix 2 — Constellation Diagrams

When to use

As already mentioned on page 10 constellation diagrams are a more sophisticated version of flow or systems diagrams. They are used to illustrate the relative strengths that exist between the relationships of the different variables within the system. Each variable must display a quantitative value so that a multiple correlation or association test can be carried out using all the variables.

Examples

Slope and soil variables, plant associations along a succession, stream flow variables.

Method of construction

As already intimated this can be very complex.

i. Carry out a multiple correlation or association test on the observed variables. List those that show a significant relationship at your chosen level of confidence (usually .05). Work out the reciprocal value for these significantly paired variables.

ii. The constellation diagram is constructed. Each variable may be put in a frame or 'box'. The length of the line (distance between each of the boxes) is proportional to the strength of the relationship between the two variables, ie. the stronger the relationship the closer the two variables are drawn next to each other. Work out a suitable scale. Variables that show no significant relationships are not connected by a line.

It is advised that you start with the one variable that appears to have the greatest number of significant relationships and work outwards. The diagram is three dimensional and as such it is very difficult to construct. Broken lines may be used to represent relationships that cannot be shown on a two dimensional graph (see worked example below).

This particular type of diagram may be rather difficult to construct but if drawn properly it is very useful in illustrating the dynamics within a system or the associations of the variables (especially useful for illustrating plant associations). Discrete populations or communities may be easily identified as seen below.

Worked example

Plant associations along a sand dune succession.
(courtesy Neil Adams — Blackburn Computer Centre)

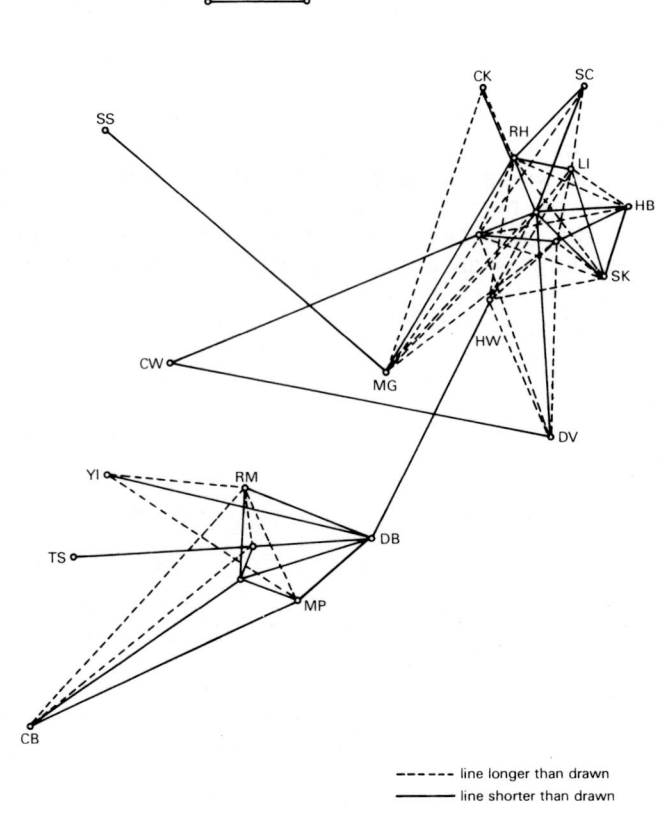

KEY TO PLANT SPECIES

MG	= Marram grass
SS	= Sand sedge
HR	= Heather
MO	= Moor grass
CK	= Chickweed
SC	= Sand couch grass
RH	= Restharrow
LI	= Lichen
HB	= Hawkbit
SK	= Seakale
HW	= Hawkweed
DV	= Dog violet
DB	= Dewberry
MP	= Marsh pennywort
RM	= Reedmace
YI	= Yellow Iris
TS	= Thistle
CB	= Cranesbill
CW	= Creeping Willow

- - - - - line longer than drawn
——— line shorter than drawn

Glossary of Words and Terms

Azimuth. A compass bearing. Often compass bearing data is put into classes or groups, these are called **azimuthal classes**.

Component categories. Elements that go to make up the total or whole, for example, a traffic count may consist of component categories such as cars, lorries, etc.

Correlation. A relationship or connection between different factors or variables.

Data. Measurements or observations usually made in the field, for example stone sizes, building ages, land uses, etc., **Continuous data** — information collected continuously over time and space. **Discontinuous data** — information collected at intervals, either regular intervals or at random.

Hypothesis. Plural hypotheses. Usually a statement that involves an idea often regarding a geographical concept, for example air temperatures will decrease with altitude. Hypotheses are often used as a framework for fieldwork enquiries. **Null hypothesis** — the converse statement, for example there is no connection between air temperatures and altitude.

Interpolation. The estimation of intermediate values and patterns using known observed data (see page 52 for example).

Orientation. The direction a phenomena (stone, corrie) faces. Usually expressed as a compass bearing.

Proportions. The relation of one variable to another, a ratio between two quantitites. These values are said to be proportional.

Qualitative. A non-numerical value or opinion, for example an environmental perception — 'a beautiful view'.

Quantitative. A measurement *or* measurable amount, for example, the length of a stone = 10.3 cms.

Standard deviation. A quantity calculated to illustrate the extent by which a group of data differs from the mean (average).

Totals. The observed data added together. These are different types: **Partial** — adding together some of the data but not all; several partial totals may go to make up the grand total. **Accumulative** — a running total as the data is systematically worked through. **Grand** — the complete end total of all the observed data.

References

As already explained in the introduction this manual has been designed to complement existing texts on fieldwork techniques. The following is therefore a list of publications that may prove interesting reading.

Chalmers, N. and Parker, P., *Fieldwork and Statistics for Ecological projects*, (O.U. project guide), F.S.C. 1986.

Dowdeswell, W.H., *Ecology — Principles and practice*, Heinemann, 1984.

Greasley, B., *Project Fieldwork*, Bell & Hyman, 1984.

Ingle Smith, D. and Stopp, P., *The River Basin*, Cambridge University Press, 1978.

Lennon, B.J. and Cleves, P.G., *Techniques and Fieldwork in Geography*, UTP, 1984.

MESU (various), *Using Computers in Fieldwork*, MESU Publications, 1988.
Various, *Science in Geography Series* (set of 4), Oxford University Press, 1970s.

Various, *Sources and Methods in Geography* (set of 6), Butterworth, 1970s.

Warn, S. and Bottomley, C., *Fieldwork Investigations* (set of 5), Arnold Wheaton, 1986.

Computer Software

Many of the techniques described in this booklet can be achieved by the use of computer software. In many cases it may be just as quick to construct the graph/map manually but the following list is included for those readers who feel that they really do need to use such technology. The list is by no means exhaustive and will soon be hopelessly out of date.

Ecosoft, AUCBE; Bar graphs, Pie graphs, Scattergraphs and Kite diagrams.
Field Study techniques, MJP; Scattergraphs, Rose diagrams, Valley profiles and River channels.
Scattergraphs, MJP; Data files and Scattergraphs.
Orientation Analysis, MJP; Rose diagrams (Vectors).
Channel Analysis, MJP; River Channels and Composite bars for sediment analysis.
Slopes, MJP; Slope profiles and proportional bars.
Project Tool Kit, MJP; Histograms, Pie charts and simple mapping.
Geobase, Longman; Comprehensive mapping program — choropleth and isopleths, cross sections and scattergraphs.

Databases often include useful mapping and graphing facilities such as Qstats, Qmap, Grass and Key, as do other commercial software packages that are designed to carry out specific functions, for example Urban Studies by SES will draw flow lines for traffic data.

Note: It is the opinion of the authors that although computers can be a useful tool in the field of data analysis students should at first attempt the required techniques manually. By manually following a method through, the user is more likely to understand the true meaning of the end result. The authors believe that merely inserting data into a machine which then in turn throws out a "picture" does not lead to a full understanding.

"Know then fully the nature of the beast so that you may be its ruler and not its slave."

THE GEOGRAPHICAL ASSOCIATION

The Geographical Association is the national subject teaching organisation for all geographers. Founded in 1893, it has over 11,000 members and 60 local branches in England, Wales and Northern Ireland. Nationally, hundreds of teachers contribute to the objectives of the Association through section committees, working groups and working parties. At a time of unprecedented change in the education system, the Geographical Association is working to provide curriculum support for teachers and safeguard and extend geography's contribution to education at all levels.

The GA offers the following services to members:

- **Primary Geographer**, a colour magazine packed with ideas for key stages 1 and 2

- **Teaching Geography**, the journal for geography teachers in secondary schools, with full colour illustrations

- **Geography**, the senior journal of the Geographical Association, designed to meet the changing needs of lecturers, teachers and students of geography

- a nationwide network of 60 branches

- books and other resources to support teachers of geography, from reception and nursery to post-16, with substantial discounts to members

- **GA News**, a quarterly newsletter sent out free to members, with information on GA activities and events

- a three-day **Annual Conference** which attracts over 1500 delegates, with free entrance to workshops, lectures and resources exhibition

- the **Fleure Library**, a unique collection of over 20,000 books and other materials covering most aspects of geography

- specialist help and information on aspects of geography and geography teaching

If you would like more information about the GA and its work, please contact:
Miss Frances Soar, Senior Administrator, The Geographical Association,
343 Fulwood Road, Sheffield S10 3BP England
Tel (0114) 267 0666 (International +44 114 267 0666)
Fax (0114) 267 0688 (International +44 114 267 0688)

Registered charity no. 313129